Wireless Networks

Series Editor

Xuemin Sherman Shen, University of Waterloo, Waterloo, Canada

The purpose of Springer's Wireless Networks book series is to establish the state of the art and set the course for future research and development in wireless communication networks. The scope of this series includes not only all aspects of wireless networks (including cellular networks, WiFi, sensor networks, and vehicular networks), but related areas such as cloud computing and big data. The series serves as a central source of references for wireless networks research and development. It aims to publish thorough and cohesive overviews on specific topics in wireless networks, as well as works that are larger in scope than survey articles and that contain more detailed background information. The series also provides coverage of advanced and timely topics worthy of monographs, contributed volumes, textbooks and handbooks.

Feng Tian • Xiangjun Xin

High-Speed Large Capacity Optical Fiber Communications

Springer

Feng Tian
School of Electronic Engineering
Beijing University of Posts and
Telecommunications
Beijing, China

Xiangjun Xin
School of Electronic Engineering
Beijing University of Posts and
Telecommunications
Beijing, China

ISSN 2366-1186 ISSN 2366-1445 (electronic)
Wireless Networks
ISBN 978-3-031-94210-5 ISBN 978-3-031-94211-2 (eBook)
https://doi.org/10.1007/978-3-031-94211-2

This Springer imprint is published by the registered company Springer Nature Switzerland AG
The registered company address is: Gewerbestrasse 11, 6330 Cham, Switzerland

Preface

With the widespread adoption of technologies such as cloud computing, streaming services, and big data analytics, the demand for higher bandwidth and faster transmission speeds has surged. However, traditional optical fiber communication systems face challenges such as signal attenuation, dispersion, and noise, calling for innovative solutions to drive the development of next-generation high-speed and large-capacity optical fiber communication systems.

In this book, we explore the key technologies of high speed large capacity optical fiber communication in depth. It covers a range of topics including multidimensional modulation and coding techniques, wavelength division multiplexing, space division multiplexing, and equalization and compensation algorithms. The detailed overview of these topics in this book provides a comprehensive understanding of the latest research and developments in fiber optic communications.

Multidimensional Modulation and Coding Technology: Encoding and modulation in optical communication is mainly used to improve the signal transmission efficiency and reliability, reducing the impact of noise and interference. Meanwhile, the multidimensional modulation and coding technologies have emerged, providing more efficient and reliable solutions for high-speed large capacity data transmission. In Chap. 2, we discuss the novel concatenated coded modulation and demonstrate its performance, such as the combination of turbo code and trellis coded modulation.

Extend Wavelength Division Multiplexing Technology: The optical fiber capacity is approaching the Shannon limit, ultra Tb/s optical transmission has become a research hotspot for high-speed large capacity optical networks, how to improve the capacity of fiber optic communication based on the existing laid fiber cables. To address this issue, multi-carrier generation and multi-band fiber transmission are introduced in Chaps. 3 and 4.

Space Division Multiplexing Technology: The space division multiplexing technology provides another new development direction for improving the capacity of optical fiber transmission systems. Through the multidimensional multiplexing of core and mode, space division multiplexing technology can increase the transmission capacity of a single fiber by tens of times on the original basis and is expected to solve the future data transmission capacity crisis. Parts of these works are presented in Chap. 5.

Equalization and Compensation Algorithm: The innovative equalization and compensation algorithms for optical fiber communication systems, aiming to improve the transmission performance in optical fiber channels and reduce inter-symbol interference caused by linear and nonlinear damage. Parts of these works are introduced in Chap. 6.

Overall, the detailed overview of these topics in this book provides a comprehensive understanding of the latest research and developments in fiber optic communications. It highlights the potential of these advancements to address challenges faced by legacy systems and to develop more efficient and reliable fiber optic communication systems. Although we made an earnest endeavor for this book, there may still be errors in the book, we would highly appreciate it if you contact us when you find any.

Beijing, China Feng Tian
Beijing, China Xiangjun Xin

Acknowledgments

Thanks to all the collaborators who have also contributed to this book: Jing Zhang, Tianze Wu, Jue Wang, Yutian Li, Chuanji Yan.

Competing Interests The authors have no competing interests to declare that are relevant to the content of this manuscript.

Contents

1 Introduction . . . 1
1.1 Introduction . . . 1
References . . . 12

2 Multidimensional Modulation and Coding Technology . . . 13
2.1 Adaptive Non-uniform PSK Modulation . . . 13
2.2 Adaptive 3-Dimensional Hierarchical QAM . . . 18
2.3 Probabilistic Shaping Four-Dimensional Modulation . . . 25
2.4 ATTCM Concatenated Coded Modulation . . . 35
2.5 Probabilistic Shaped Trellis Coded Modulation . . . 40
2.6 Summary . . . 48
References . . . 48

3 Multi-carrier Generation and Transmission Technology . . . 51
3.1 Theoretical Analysis . . . 52
3.2 Simulation and Experimental Results . . . 60
3.3 RFS-Generated 20-Carrier Optical Comb for PDM-RZ-16QAM CO-WDM . . . 73
3.4 Summary . . . 77
References . . . 77

4 Multi-band Fiber Transmission (MBT) Technology . . . 81
4.1 Fast-Than-Nyquist Rate (FTN) Technology . . . 84
4.2 Performance Analysis and Power Tilt Mitigation of Ultra-Wideband WDM Transmission Systems . . . 89
4.3 Ultra-Wideband Transmission . . . 104
4.4 Summary . . . 108
References . . . 109

5 Space Division Multiplexing (SDM) Technology . . . 113
5.1 Basic Principle . . . 113
5.2 SDM Transmission System . . . 117

5.3 Experimental with XT-MIMO-Volterra Algorithm Over a 7-Core Fiber with a KK Receiver 125
5.4 MCF-SHCD System Based on PS MD Modulation 141
5.5 Summary 148
References 149

6 Channel and Equalization Algorithms in Optical Fiber Communication System 153
6.1 Equalization and Compensation Algorithm 153
6.2 Jones-Volterra Training Sequence (JVTS) Algorithm for Nonlinear Compensation 159
6.3 Carrier Phase Recovery Algorithm 163
6.4 Summary 178
References 179

7 Conclusions and Future Challenges 183
7.1 Conclusions and Future Challenges 183
7.2 Future Challenges 184
7.3 Summary 185

Chapter 1
Introduction

1.1 Introduction

1.1.1 Background

In today's hyper-connected world, where information flows constantly and seamlessly, the significance of high-speed large capacity optical fiber communication systems cannot be overstated. These systems have become the lifeblood of our global communication networks, enabling the exchange of vast amounts of data across vast distances with unprecedented speed and reliability [1]. Optical fibers, being the fundamental building blocks of these systems, have revolutionized the telecommunication industry, transforming the way we live, work, and interacting with each other.

The demand for high-speed large capacity optical fiber communication systems has grown exponentially in recent years, driven by the increasing popularity of cloud computing, streaming services, and big data analytics [2]. As these technologies become more widespread, the need for more bandwidth and faster data transmission speeds has become urgent. This has led to a surge in research and development efforts aimed at pushing the boundaries of optical fiber communication technology to meet these demands [3].

Moreover, with the rapid advancements of technology, the traditional optical fiber communication systems are facing great challenges. The limitations of existing technologies, such as signal attenuation, dispersion, and noise, are becoming increasingly apparent [4, 5]. This calls for innovative solutions that can address these challenges and enable the development of next-generation optical fiber communication systems with even higher speeds and capacities.

In this context, the book "High-speed Large Capacity Optical Fiber Communications" explores the principles and applications of high-speed large capacity optical fiber communication in depth. It covers a range of topics including multi-dimensional modulation and coding techniques, multi-band fiber transmission,

F. Tian, X. Xin, *High-Speed Large Capacity Optical Fiber Communications*,
Wireless Networks, https://doi.org/10.1007/978-3-031-94211-2_1

space division multiplexing, and equalization and compensation algorithms. A detailed overview of these topics in this book provides a comprehensive understanding of the latest research and developments in fiber optic communications. It highlights the potential of these advancements to address challenges faced by legacy systems and to develop more efficient and reliable fiber optic communication systems.

1.1.2 Overview of Related Technologies

In this section, we will introduce the technologies of optical fiber communication concerned.

1.1.2.1 Probabilistic Shaping

Probabilistic Shaping (PS) is an optimization technique for modulation formats that provides very high flexibility for optical communication without increasing system complexity. PS technology has been widely used in optical fiber communication systems to narrow the gap and to achieve high transmission capacity with lower computation complexity, and can also overcome power limits caused by nonlinearity as well [6].

For conventional MQAM modulation, the amount of information carried by each symbol is an integer. So, when designing a given line rate, such as 200G, if double-biased QPSK modulation is being used, the baud rate will reach about 64G; and if double-biased 8QAM modulation is being used, the baud rate will be up to about 45G and thus the adjustment of the baud rate will be very limited. If the Flex grid of 12.5 GHz is considered to be appropriate, the channel spacing of at least 75 GHz and 62.5 GHz is required, which is unfavorable for improving spectral efficiency. To ensure higher spectrum utilization efficiency, the baud rate should be adjusted to fill the smallest channel interval possible [7]. For example, a 200G signal is filled within a 50 GHz interval. You might think that with 16QAM, only 32Gbaud can be transmitted in the 50 GHz interval. However, the transmission distance of 200G 16QAM can be only within 600 km, which cannot meet many long-distance backbone network applications. Therefore, QAM modulation not only wastes SNR resources, but also brings a lot of trouble to system design and optimization. Moreover, it can only achieve discrete distance and capacity, and thus cannot cover all scenes continuously and efficiently.

How to solve it? It is proposed that by changing the probability of symbols appearing in different positions of the constellation diagram, the frequency of the outer constellation points is reduced, which is conducive to reducing the average power, equivalent to increasing the minimum Euclidean distance, so as to present better transmission performance. This is what we call probabilistic shaping (PS) [8].

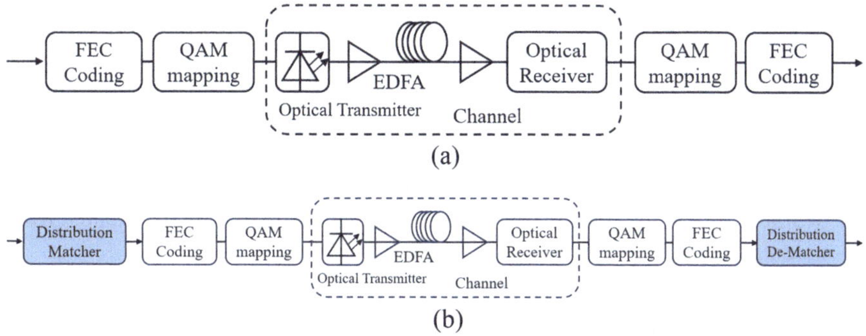

Fig. 1.1 (**a**) Communication system block diagram. (**b**) Add PS communication system block diagram

PS is implemented by adding a Distribution Matcher (DM) before the encoding and a Distribution De-Matcher (de-matcher) after the decoder, as shown in Fig. 1.1.

Taking 16QAM as an example, the probability of each symbol appearing in the obtained 16QAM signal through the combination of out-of-matcher coding and FEC coding follows our preset probability distribution, and this probability distribution is determined by SNR. It increases the probability of four constellation points in the inner circle and decreases the probability of our constellation points in the outer circle, thus improving the error performance and saving the transmission power. Generally, the worse the SNR, the greater the degree of probability shaping. In theory, each SNR has a corresponding optimal probability distribution, which can help the system achieve the best optimization [9].

Why does probabilistic shaping improve error performance? The 16 constellation points of 16QAM have three different magnitudes, namely three circles. The distance between the inner circle and the second circle is 1.75, and the distance between the second circle and the outer circle is 1.08. So, the outer ring only needs to be wrong 1.08 distance that will lead to no code, but the inner ring needs to be wrong 1.75 distance that will lead to no code. Therefore, using 16QAM as a modulation format, the biggest contribution to no code is the four constellation points in the outer circle. Therefore, probability shaping reduces the probability of constellation points in the outer circle and increases the probability of constellation points in the inner circle, thereby reducing the bit error rate [10]. On the other hand, reducing the probability of constellation points in the outer circle and increasing the probability of constellation points in the inner circle mean that the average power of the signal after probability shaping is lower than that of the original signal, which is conducive to avoiding the power limitation problem caused by the nonlinear problem of optical communication. In order to compare the performance of the two, we need to compare under the same transmission power and expand the constellation map after probability modification, which means that the Euclidean distance between constellation points is improved, and the fault tolerance is improve [11]. With the increasing of Euclidean distance, the bit error rate will decrease under the same SNR.

In summary, the advantages of probabilistic shaping include:

(a) It has a shaping gain and is close to the Shannon limit.
(b) It is expected to achieve higher transmission capacity and significantly improve spectrum efficiency.
(c) The transmission rate can be flexibly adjusted to perfectly adapt to different transmission channels.
(d) Multiple QAM mappings are not needed, except the use of square QAM modulation and the need to adjust the shaping coefficient.
(e) No need to adjust FEC and other technical solutions

1.1.2.2 Wave-Band Extension

Since the commercialization of DWDM technology, long-distance systems have continuously expanded the optical fiber transmission bandwidth: from the early C-band (C4T) to the CE-band (C4.8T), and then to the C++-band (C6T) [12]. The commercial landing of 80-wave 75 GHz spaced 200 Gbit/s QPSK or 120-wave 50 GHz spaced 200 Gbit/s 8QAM/PS16QAM schemes has increased the single-fiber capacity by 50%. In fact, the low-loss window of single-mode fiber not only includes the C-band, but also contains the O, E, S, L, U and other bands. Among them, the L-band has a small amount of deployment among Japanese operators. The deployment of the L-band can avoid the impact of four-wave mixing in G653 dispersion-shifted fiber. In recent years, a small number of operators and Internet vendors in the United States have deployed C + L systems in DCI and submarine cable transmission, which can double the fiber capacity. As single-mode fiber approaches the Shannon limit of 100 Tbit/s in capacity, band expansion technology has become a research hotspot in both academic and industry circles [13].

Extended C-band: The expansion plan for the C-band can be divided into two stages. The first stage is to expand the 80-wave system with 50 GHz spacing (C80) to 96 waves (C96), increasing the system capacity by 20%. With the increase of traffic and the development of technology, the second stage of the expansion plan aims to, extend to a 120-wave system with 50 GHz spacing, i.e., the expansion of the C-band. Compared to the 80-wave system, the system capacity increases by 50%. Currently, the extended C-band with an 80-wave PM-QPSK wavelength division system operating at a wavelength spacing of 75 GHz is mature, and equipment from major equipment vendors can support it.

Compared to C80 and C96, the expanded C-band system needs addressing a series of technical issues, including optimizing the design of line cards, increasing the wavelength tuning range of lasers, optimizing optical amplifiers to meet the amplification gain and flatness requirements of the expanded C-band, and optimizing wavelength-selective switches (WSS) to meet the application requirements of the expanded C-band. The application of the expanded C-band can introduce a certain degree of degradation in transmission performance, resulting in an overall system performance degradation of 0.5 to 1 dB compared to C80/C96 systems. Via the relevant system optimization design, the degradation can be limited within 0.5 dB. In addition, the key components of the expanded C-band WDM system need to be further optimized and improved for better performance. Currently, WDM there is still a lack of relevant system parameter standard definitions. It is necessary to accelerate the upgrade of relevant standards to promote industrial development.

C + L band: As a mature commercial technology, C + L has been widely used in Japan, Europe, and other countries and regions, effectively doubling the transmission capacity of systems (80 × 2 waves or 96 × 2 waves). In areas with limited fiber resources and high laying or leasing costs, it is an appealing option. In China, the CCSA has conducted relevant research on L-band 100 Gbit/s WDM systems. Due to the wide spectral characteristics of C + L systems, higher standards are placed on dynamic balancing of channel optical power, nonlinear compensation, and other aspects. Because the optical fiber attenuation in the L-band is greater than that in the C-band, the OSNR in the L-band is degraded by 0.5–1 dB compared to the C-band. In addition, the presence of stimulated Raman scattering effects can cause power transfer from the C-band to the L-band, leading to degradation in C-band performance. These issues need to be carefully considered during the research and application design of C + L band.

Other band: Compared to the C and L bands, the commercial applications of the O, E, S, and U bands are relatively limited and still in the exploration and research stage. As one of the earliest bands used, the O-band has a certain technological foundation, but its application in large-scale high-speed transmission scenarios is limited by its loss characteristics and other issues in long-distance transmission. E. The S and U bands face more complex technical challenges due to compatibility with existing communication systems and the progress of related optical device development, including the adaptation of optical amplifiers, device design and manufacturing, etc. However, these bands provide potential directions for further expansion of fiber optic transmission bandwidth in the future. With the continuous development of communication technology, bottlenecks in related technologies have been overcome. At present, the research team of the National Institute of Information and Communications Technology (NICT) in Japan has developed an optical transmission system that achieves a data transmission rate of 402 Tb/s in standard commercial optical fibers by combining various new amplification technologies. The system covers all transmission bands (O, E, S, C, L, U) within the low loss window of standard optical fibers and achieves a transmission bandwidth of 37.6 THz. The O, E, S, and U bands are expected to play an important role in future communication networks, providing new solutions to meet the growing demand for communication capacity.

1.1.2.3 Equalization Algorithm

In high-speed large capacity optical fiber communication systems, the implementation of equalization algorithms is indispensable for mitigating complex channel impairments and enhancing overall system performance. With the exponential growth of global data traffic, driven by advancements in cloud computing, 5G networks, the Internet of Things (IoT), and artificial intelligence (AI), optical fiber communication systems are increasingly required to support higher transmission rates (e.g., Tb/s levels) and longer transmission distances (ranging from hundreds to thousands of kilometers). However, the optical fiber channel is subject to a variety of physical impairments that significantly degrade signal quality and limit system performance. Consequently, equalization algorithms, leveraging digital signal processing (DSP) techniques, have emerged as a critical solution to dynamically compensate for these impairments and ensure reliable signal transmission.

Linear impairments are among the most significant challenges in optical fiber communication systems. Chromatic dispersion (CD) and polarization mode dispersion (PMD) are two primary forms of linear impairment. CD causes temporal broadening of optical pulses, leading to inter-symbol interference (ISI), which becomes increasingly severe at higher data rates due to the accumulation of dispersion effects. PMD, on the other hand, results in differential group delays between orthogonal polarization states, causing signal distortion. Additionally, polarization-dependent loss (PDL) introduces random fluctuations in the polarization state of the signal, further degrading the performance of polarization-division multiplexing (PDM) systems. These linear impairments necessitate the use of advanced equalization algorithms, such as minimum mean square error (MMSE) equalizers, to mitigate their impact and restore signal integrity.

Nonlinear effects in optical fibers pose another major challenge, particularly in long-haul, high-power transmission systems. The Kerr effect, including self-phase modulation (SPM), cross-phase modulation (XPM), and four-wave mixing (FWM), interacts with dispersion to cause irreversible signal distortion. Furthermore, stimulated scattering effects, such as stimulated Brillouin scattering (SBS) and stimulated Raman scattering (SRS), exacerbate power imbalances within the channel. These nonlinear impairments are particularly pronounced in large capacity systems, where traditional fixed compensation methods, such as dispersion-compensating fibers (DCFs), are insufficient. To address these issues, nonlinear equalization techniques, including Volterra series-based approaches and machine learning algorithms, are employed to model and compensate for nonlinear distortions dynamically.

In addition to fiber-related impairments, device imperfections and noise sources within the communication system also contribute to signal degradation. For instance, phase noise in laser sources introduces carrier phase jitter, leading to constellation rotation and demodulation errors in coherent optical communication systems. Moreover, the nonlinear response of electro-optic modulators and photodetectors can introduce harmonic distortion, reducing signal fidelity. These device-related impairments require sophisticated equalization algorithms to compensate for their effects and ensure high-quality signal transmission. The advent of space-division multiplexing (SDM) technologies, such as few-mode fibers (FMFs) and multi-core fibers (MCFs), has further increased the complexity of optical fiber channels. In FMFs, mode coupling introduces mode-dependent loss (MDL) and mode crosstalk, while in MCFs, inter-core crosstalk results in signal interference between adjacent cores. These impairments necessitate the use of specialized equalization techniques, such as multiple-input multiple-output (MIMO) equalization, to separate and recover signals from different modes or cores effectively.

In conclusion, equalization algorithms are essential for overcoming the multifaceted challenges posed by linear and nonlinear impairments, device imperfections, and SDM-related crosstalk in high-speed large capacity optical fiber communication systems. By leveraging advanced DSP techniques, these algorithms enable real-time estimation and compensation of channel impairments, significantly enhancing system robustness and performance. As optical communication systems continue to evolve toward higher data rates and greater complexity, equalization algorithms will remain a cornerstone technology, enabling the realization of ultra-100 Gb/s coherent transmission, SDM, and future Pbit/s-class optical networks. Their adaptive and flexible nature ensures that they will continue to play a critical role in optimizing the performance and capacity of next-generation optical communication systems.

Linear equalization algorithm: The fundamental principle of linear equalization algorithms is to process the received signal by constructing a linear filter to eliminate distortions generated during channel transmission. This linear filter can adaptively adjust based on inter-symbol interference (ISI) caused by channel characteristics such as dispersion, polarization mode dispersion, frequency response of devices. Channel characteristics to achieve optimal signal compensation. In practical applications, linear equalization algorithms are often combined with other signal processing techniques, such as modulation/demodulation and encoding/decoding, to achieve efficient and reliable signal transmission.

In recent years, researchers have conducted extensive work on optimizing and improving linear equalization algorithms. Among them, the Least Mean Square (LMS) algorithm and Recursive Least Squares (RLS) algorithm are two representative optimization methods. The LMS algorithm is a gradient descent-based optimization algorithm that continuously adjusts filter coefficients to minimize the mean square error between the output signal and the desired signal. This algorithm has the advantages of simplicity and ease of implementation, yet bears limitations in terms of convergence speed and stability. On the other hand, the RLS algorithm is based on recursive least squares method, which recursively computes filter coefficients to minimize the sum of squared errors between the output signal and the desired signal. Compared to the LMS algorithm, the RLS algorithm has faster convergence speed and better stability yet relatively higher computational complexity. In addition to the LMS and RLS algorithms, researchers have proposed many other optimization methods, such as linear equalization algorithms based on neural networks and particle swarm optimization. These methods have their own characteristics and to some extent, and can improve the performance of linear equalization algorithms. With the ceaseless development of computer technology, the implementation of linear equalization algorithms has become more efficient and precise, providing strong support for practical applications.

Nonlinear equalization algorithm: In fields such as optical fiber communication, various nonlinear factors, such as the Kerr effect and Raman scattering in optical fibers, affect signal transmission, leading to signal distortion and performance degradation. Traditional linear equalization algorithms often struggle to address these nonlinear distortion problems, thus necessitating the use of nonlinear equalization algorithms to effectively mitigate such effects.

The fundamental idea behind nonlinear equalization algorithms is to construct more complex mathematical models to approximate the nonlinear characteristics of the channel, thereby accurately restraining signal distortion during transmission. This includes utilizing machine learning techniques such as neural networks and support vector machines (SVMs) to train and learn the nonlinear characteristics of the channel, enabling real-time equalization of signals during actual transmission. In the application of neural networks, nonlinear equalization algorithms can simulate the nonlinear behavior of the channel by constructing deep neural networks. With extensive training data, neural networks can learn the intricate characteristics of the channel and generate corresponding equalizer parameters. During actual transmission, these parameters are used to process received signals in real-time, effectively eliminating nonlinear distortion. SVMs, on the other hand, are machine learning algorithms based on statistical learning theory, which excel in handling high-dimensional data and nonlinear problems. In nonlinear equalization algorithms, SVMs can be used to construct classifiers or regression models to approximate the nonlinear characteristics of the channel. By selecting appropriate kernel functions and parameter optimization methods, SVMs can secure the optimal decision boundary or regression function in high-dimensional feature space, thereby effectively compensating for nonlinear distortion. In addition to neural networks and SVMs, nonlinear equalization algorithms can also draw inspiration from other machine learning algorithms such as decision trees and random forests. These algorithms have their own strengths and characteristics when dealing with different types of data and problems, providing nonlinear equalization algorithms with a more diverse and rich set of tools.

1.1.3 Related Research and Development

1.1.3.1 400G QPSK

To advance the deployment of computational networks and the "Eastward Shift, Westward Calculation" layout, it is necessary to promote research and development in key technologies for 400G, achieving comprehensive technological innovation in modulation, spectrum, and infrastructure.

As early as 2018, China Mobile, in collaboration with equipment vendors, conducted tests in live networks using single-carrier 400G 16QAM and achieved a transmission distance of up to 600 km. In October 2021, China Mobile, together with Huawei, ZTE, and Fiber Home, among other equipment vendors, completed the world's first ultra-wide spectrum single-wave 400G large capacity optical transmission verification in live networks, achieving a transmission distance of over 1000 km. In July 2022, China Mobile, in collaboration with ZTE Corporation, conducted laboratory simulations of fiber length, loss, and maintenance margin based on live network requirements and verified 400G QPSK transmission, achieving a transmission distance of 3038 km across 49 spans without electrical relay. Based on the laboratory test results, in January 2023, China Mobile conducted live network testing of 400G QPSK, spanning four provinces including Zhejiang, Jiangxi, Hunan, and Guizhou, involving 45 optical amplification sections. It achieved a record-breaking 5616 km ultra-long-distance land-based real-time live network transmission using 400G QPSK without electrical relay, and for the first time, the transmission performance of spectrum expansion to 12THz in the C6T + L6T band is verified.

1.1.3.2 800G OTN

In the digital era, with the widespread application of technologies such as cloud computing, big data, and AI, data traffic is experiencing exponential growth, calling for higher demands on network bandwidth. To cope with the rapid growth of traffic, the port rates of OTN transmission networks are continuously upgraded: from the commercialization of 100G rates in 2012, to the large-scale deployment of 200G wavelength division in 2016, and the emergence of 400G wavelength division in 2020. The surge in business traffic has driven the updating and iteration of transmission rates. Currently, there is a growing demand for 800G OTN from operators.

ZTE Corporation has introduced the industry's first low-power pluggable 800G OTN solution, which can help operators reduce the cost and power consumption per Gbit, thus creating a more environmentally-friendly and more efficient network infrastructure. In December 2023, ZTE Corporation completed the world's first on-site verification of a computational interconnection solution based on 800G 16QAM pluggable modules in China. In February 2024, ZTE Corporation completed single-wave 800G non-electrical relay transmission verification in Turkey based on 800G pluggable coherent optical modules, achieving a transmission distance of up to 2000 km, the farthest for land systems.

ZTE Corporation has launched the industry's first 800G pluggable OTN solution, boasting four major advantages: ultra-flexibility, high performance, large capacity, and comprehensive applicability.

Ultra-flexibility: Firstly, the pluggable modules are considerably smaller, reducing volume by 50% compared to fixed modules and thereby minimizing space occupation while doubling the port density per board. Secondly, the pluggable modules exhibit a 68% reduction in Gbit power consumption compared to fixed modules, substantially lowering the overall system energy consumption. Pluggable optical modules, featuring hot-swappable capabilities, facilitate on-demand and rapid capacity expansion, thus significantly saving customer CapEx.

High performance: ZTE's 800G pluggable solution Based on 130GBd technology, ZTE's 800G pluggable solution can deliver greatly enhanced transmission performance compared to 96GBd fixed modules. At 400G speed, transmission distance increases by 60%, while at 800G speed, transmission distance doubles.

Large capacity: In terms of single-wave rate, ZTE has extensive practical experience with 1.2 T and continues to explore higher 1.6 T capacities. In spectrum expansion, ZTE has extended commercial spectrum to the C + L band, providing 12Thz spectrum resources, and is further expanding to the S + C + L band. ZTE has already conducted Tbit-level real-time transmission system tests based on the S + C + L band, achieving a maximum laboratory single-fiber capacity of 136 T, and will continue to explore even larger capacity OTN transmission systems.

Comprehensive applicability: ZTE's full range of OTN product solutions, based on 800G pluggable optical modules, covers backbone, metropolitan, and DCI scenarios. For backbone scenarios, the farthest transmission distance can be achieved by employing a 400G QPSK modulation scheme based on 130GBd per wave. For metropolitan scenarios, depending on the transmission distance, either a 400G scheme with a 75 GHz/100 GHz channel spacing or an 800G scheme with a 150 GHz channel spacing can be adopted to achieve the optimal balance between transmission distance and capacity. For DCI scenarios, the maximum system capacity can be realized by adopting an 800G 16QAM modulation scheme based on 130GBd per wave.

1.1.4 Structures of the Book

This book aims to offer a thorough understanding of the fundamental technologies and advancements in high-speed large capacity optical transmission systems. Each chapter builds upon its predecessors, facilitating a gradual and progressive exploration of the field.

Chapter 1 lays the groundwork by introducing the background. It sets the stage for the subsequent in-depth discussions by presenting an overview of the high-speed large capacity optical transmission system, highlighting its key technologies and challenges. Additionally, it provides a sneak peek into the related research and development efforts in the field and outlines the overall structure of the book.

In Chap. 2, the spotlight shifts to cutting-edge modulation and coding strategies designed for next-generation elastic optical networks. It commences with an adaptive non-uniform PSK modulation scheme, which intelligently tailor's constellation amplitudes to meet specific quality of service requirements while maintaining receiver simplicity. This is followed by the introduction of a 3D hierarchical QAM framework that exploits multi-dimensional constellation reconfiguration for enhanced error resilience. The chapter further explores a probabilistic shaping 4D modulation scheme within a self-homodyne coherent transmission system to boost spectral efficiency. It concludes with a hybrid concatenated coding scheme in a GFDM architecture, combining adaptive turbo codes and trellis-coded modulation (ATTCM) to achieve robust coding gains with minimal complexity. We further propose a novel probabilistic shaping trellis-coded modulation (PS-TCM) scheme integrated with GFDM in an intensity-modulation/direct detection (IM/DD) system.

Chapter 3 delves into the progress of multi-carrier generation and large-capacity optical transmission systems enabled by recirculating frequency shifter (RFS) technology. It begins by describing how an SSB modulator-based RFS is employed to generate a large number of frequency-locked optical carriers with outstanding flatness and stability, while also experimentally investigating the factors influencing carrier flatness. Subsequently, a novel dual-band RFS architecture is proposed, doubling the efficiency and halving the costs. The chapter culminates in the experimental verification of a 2.56-Tb/s PDM-RZ-16QAM CO-WDM system over 800 km of standard single-mode fiber using RFS-generated optical combs, achieving a remarkable spectral efficiency.

Chapter 4 explores ultra-wideband (UWB) wavelength division multiplexing (WDM), to augment the capacity of coherent fiber optic communication systems. It starts with the presentation of a new multi-band transmission scheme incorporating the Fast-than-Nyquist rate and a non-orthogonal technique. Then, it dives into the analysis of a UWB WDM system spanning multiple bands, including the construction of the system and a study of channel effects. An optimization plan combining a Raman amplifier and partition particle swarm optimization (PPSO) to correct spectral tilt is also detailed. The chapter concludes with the achievement of ultra-wideband transmission and a high data rate using an artificial neural network to combat fiber nonlinearities.

Chapter 5 centers on SDM, beginning with a discussion of its basic principle and key technology. It then proposes the VSS-FDLMS algorithm and demonstrates its application in a 7-core 6-mode optical SDM transmission system. The chapter further showcases 16-QAM signal transmission with a KK receiver and the employment of MD modulation formats for the MCF-SHCD system.

Chapter 6 is dedicated to novel equalization and compensation algorithms for optical fiber communication systems. It first presents a joint CMA-DDLMS algorithm that enables efficient 2 × 20G Baud PAM4 transmission over 20 km, reducing bit error rate and analyzing relevant factors. And demonstrated a SCA-CMMA-based MIMO equalization method Next, the Jones-Volterra training sequence (JVTS) is introduced to mitigate nonlinearities and crosstalk in fiber mode field (FMF) systems, improving received power and bit error rate. The chapter wraps up with a feed-forward carrier phase estimation algorithm that reduces complexity while ensuring accuracy.

Finally, Chap. 7 summarizes the key technologies explored throughout the book in the realm of high-speed large capacity optical fiber communication. It also takes a forward-looking approach, analyzing the future challenges that lie ahead in this rapidly evolving field, providing a comprehensive conclusion and a springboard for further research.

1.1.5 Summary

In this chapter, we mainly introduce the background. We first propose a high-speed large capacity optical transmission system and discuss the key technologies and challenges of this system. Then, the related research and development are provided. Finally, we give the structure of this book.

References

1. S.H. Yu, W. He, A comprehensive review of the development of optical fiber communication technology. Sci. China Inf. Sci. **50**(9), 1361–1376 (2020)
2. S.H. Yu, X.Z. Hu, *Research on the Leading Edge of Ultra High Speed, Ultra Large Capacity and Ultra Long Distance Optical Fiber Transmission System* (Science Press, Beijing, 2014)
3. P.J. Winzer, Scaling optical fiber networks: Challenges and solutions. Opt Photon. News **26**, 28–35 (2015)
4. S.H. Yu, Ten characteristics of network communication technology. People's Posts Telecommun. News (005), 8–16 (2018)
5. S.H. Yu, The seven technical walls of network communications and the primary exploration of trends. Study Opt. Commun. **44**, 1–7 (2018)
6. X. Liu, J. Zhang, M. Zhu, 光通信中概率整形技术的研究现状和进展. Laser Optoelectron. Prog. **59**(11), 1100004 (2022)
7. L. Wang, X. Chen, X. Mao, et al., Performance improvement of coherent optical chaos communication using probabilistic shaping [J]. Opt. Lett. **48**(4), 1008–1011 (2023)
8. B. Liu, L. Jiang, Y.Y. Mao, et al., Probabilistic shaping coded modulation technology [I]. J. Nanjing Univ. Inform. Technol. (Nat. Sci. Ed.) **12**(1), 10 (2020)
9. K. Wu, J. He, Z. Zhou, et al., Probabilistic amplitude shaping for a 64-QAM OFDM W-band RoF system. IEEE Photon. Technol. Lett. **31**(13), 1076–1079 (2019)
10. S. Chester, Z. Qiulin, Probabilistic shaping: A step closer to the Shannon limit in channel capacity for optical communications. Electr. Lett. (Wiley-Blackwell) **56**(22), 1162–1163 (2020)
11. S. Zhang, Z. Qu, F. Yaman, et al., Flex-rate transmission using hybrid probabilistic and geometric shaped 32QAM, in *Optical Fiber Communication Conference*, (Optica Publishing Group, 2018), p. M1G.3
12. ITU-T. Optical Fibers, Cables and System [EB/OL]. (2009-11-28)
13. A. Ferrari et al., Assessment on the achieveable throughput of multi-band ITU-T G.652.D fiber transmission system. J. Lightwave Technol. **38**(16), 4279–4291 (2020)

Chapter 2
Multidimensional Modulation and Coding Technology

In the field of communication, Multidimensional Modulation and Coding Technology have gained great attention in recent years. With the ever-increasing demand for higher data transmission rates, traditional modulation and coding methods have become insufficient for modern communication systems. Therefore, Multidimensional Modulation and Coding Technology have emerged, providing more efficient and reliable solutions for data transmission.

Multidimensional modulation technology leverage information from multiple dimensions for modulation, increasing the information capacity of the signal in multiple dimensions. Coding techniques, on the other hand, are developed to enhance the reliability of data transmission. By adding redundant information to the data, coding techniques can correct errors generated during transmission, ensuring correct data reception. With the combination of multidimensional modulation and coding techniques, higher data transmission rates and lower bit error rates can be realized. Therefore, this subsection will introduce the principles, applications and development trends of multidimensional modulation and coding techniques to provide readers with a reference for understanding the research results and development trends in this field.

2.1 Adaptive Non-uniform PSK Modulation

With the explosive development of the emerging multimedia services such as interactive games, social networks and telemedicine, different requirements have been put forward regarding the services quality provided by the optical network. In order to satisfy multi-service transmission in different conditions, non-uniform modulation techniques can be employed, which makes use of the unequal Euclidean distances between its different levels of the constellation. Since different constellations have their own bit error rate and tolerance for optical signal-to-noise ratio

F. Tian, X. Xin, *High-Speed Large Capacity Optical Fiber Communications*,
Wireless Networks, https://doi.org/10.1007/978-3-031-94211-2_2

(OSNR), the constellation spacing could be dynamically adjusted according to the channel condition and error rate requirements, so as to satisfy the demanded QoS.

In the literature, many adaptive transmission and non-uniform quadrature amplitude modulation (QAM), PSK and pulse amplitude modulation (PAM) modulation schemes have been proposed. The authors in [1] proposed a flexible 2/4/8 PAM modulation scheme for ONUs with different received optical power. This scheme used three constellation sizes to decrease the receiver complexity. An adaptive 4–64 QAM transmission was reported in [2], and the modulation was changed based on specific channel thresholds lacking flexibility. An asymmetric 8PSK bit-mappings scheme and some bounds on the BER in an AWGN channel was introduced in [3]. The authors in [4] proposed an adaptive hierarchical PSK to transmit various traffic with different QoS requirements.

In this subsection, we introduce a novel adaptive non-uniform 8PSK + 2ASK modulation scheme. Compared with quadrature amplitude modulation (QAM), PSK modulation can meet the nonlinear requirements of the system for its relatively constant envelope. Utilizing only one constellation with uniform phases, the proposed PSK scheme can effectively reduce the receiver sensitivity when compared with the traditional adaptive transmission methods [1, 2]. In addition, the continuous changes of the shape of the 8PSK + 2PSK constellation can bring flexibility in the modulation. Moreover, we propose an innovative approach to adjust the constellation envelope spacing based on the channel condition and error requirements to achieve flexible signal transmission. In this way, the receiver only needs to additionally detect simple amplitude instead of complex phase, which further reduces its complexity. For an 8PSK + 2ASK signal, duo to the unequal Euclidean distance, some bits have better tolerance for OSNR, which can be used to carry higher QoS messages, and the rest instead transmit lower QoS messages or additional messages.

2.1.1 The Principle of Adaptive Non-uniform 8PSK + 2ASK Modulation

An 8PSK + 2ASK scheme is designed based on a uniform 8PSK constellation. Specifically, every point in a uniform 8PSK constellation is split into two points, each carrying a half bit, which is named half-point. One of the half-point stays, and the other translates Δd inward in the phase direction. The two half-point in this constellation construction are used to convey less important messages or additional messages, and their relative value are controlled by Δd. Figure 2.1 illustrates the constellation construction.

The eight constellation points represented by the red circle in the 8PSK + 2ASK constellation (as shown in Fig. 2.1c) are named outer-8PSK, and the two points in the red circle are named inner-2ASK.

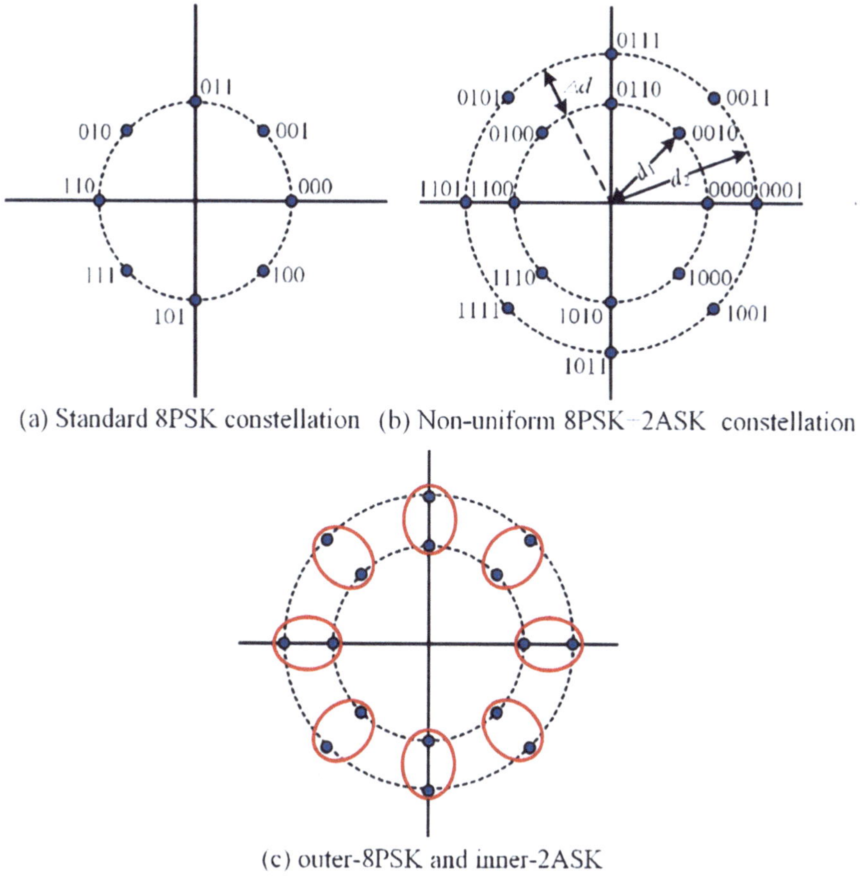

Fig. 2.1 The structure of the 8PSK + 2ASK constellation

We define the set $D = [d_1, d_2]$ that represents the envelope amplitude or the distance between constellation points and the origin. The ratio r between the envelope amplitude is given by

$$r = \frac{d_1}{d_2}, r \leq 1 \tag{2.1}$$

Where $r \leq 1$ is used to ensure that the average optical power of the signal does not exceed that of uniform 8PSK.

An 8PSK + 2ASK signal can be expressed in form by

$$s(t) = (1 + r)A\cos(w_c t + \theta) \tag{2.2}$$

Where A is the amplitude of the outer envelope, i.e. d_2. w_c is the carrier frequency, θ is the phase offset determining the distribution of constellation points, and $\theta = k\pi/7$, $0 \le k \le 7$.

In the 8PSK + 2ASK modulation, different modulation levels have different bit error rates and packet error rates (PER). The rate of correct packet reception is:

$$P_n = 1 - PER \tag{2.3}$$

Where $n = 1, 2$ presents outer-8PSK and inner-2ASK respectively. As the packets of original data stream are divided into outer-8PSK and inner-2ASK signals in 3/4 and 1/4 ratio, the transmission efficiency can be expressed as

$$\xi_n = P_n N \log_2 M \tag{2.4}$$

for N = [3/4, 1/4], where M denotes constellation points. Therefore, the overall transmission efficiency of the system is obtained by:

$$\xi_{sys} = \sum_{n=1}^{2} \xi_n \tag{2.5}$$

2.1.2 Simulation and Results

The block diagram of 8PSK + 2ASK modulation system is illustrated in Fig. 2.2. At the transmitter, a set of random sequences is transformed into four groups of data streams by serial-to-parallel conversion and mapped to 8PSK + 2ASK signals, where

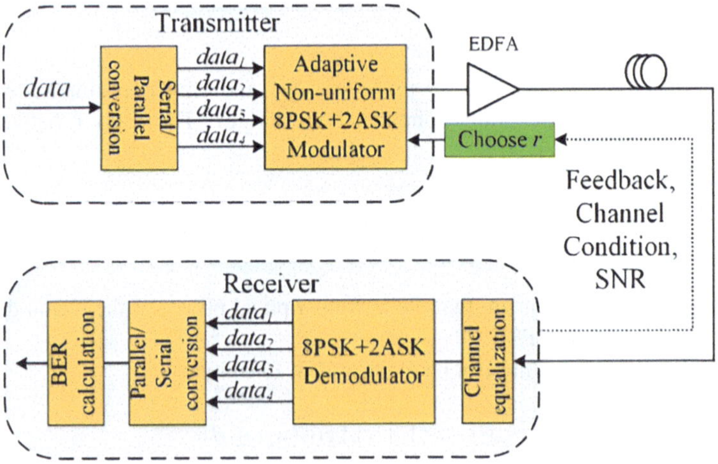

Fig. 2.2 The 8PSK + 2ASK modulation system

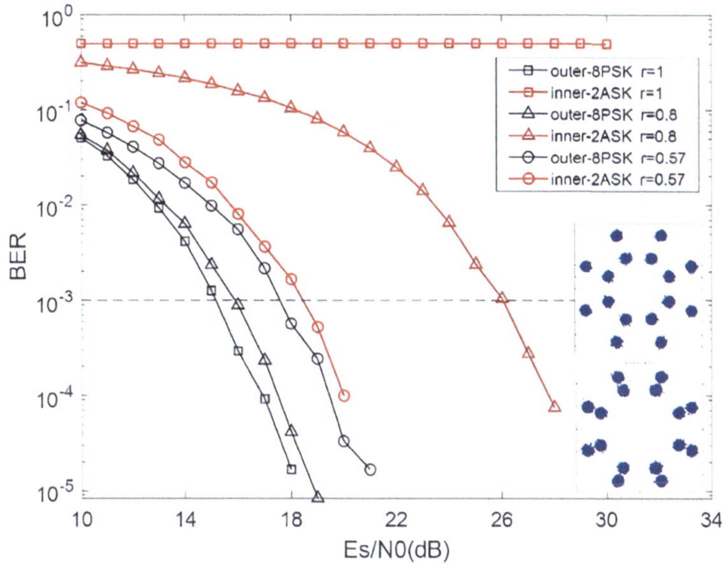

Fig. 2.3 BER performance of the 8PSK + 2ASK modulation

three groups are mapped to outer-8PSK signals and one group is mapped to inner-2ASK signals. After passing through the Gaussian white noise channel, the signals are received by the coherent receiver. With demodulation and parallel-series transformation, the BER are calculated. At the same time, the receiver continuously monitors the channel condition and feeds it back to the transmitter via the feedback signal. Then the transmitter selects the value of the envelope ratio based on feedback to meet the BER requirements for different message. When the channel condition is poor, we can increase the value of r thus preferentially transmit outer-8PSK at the cost of reducing the BER performance of inner-2ASK, or even transfer outer-8PSK signals only. When the channel condition improves, all points can be assigned with data streams, and the BER performance of inner-2ASK will also be improved.

The BER performance of the non-uniform 8PSK + 2ASK modulation is shown in Fig. 2.3. At the threshold of $BER = 1 * 10^{-}3$ (FEC limitation), when $r = 0.57$, the required SNRs for outer-8PSK and inner-2ASK are 17.5 dB and 18.4 dB respectively. When $r = 0.8$, the corresponding SNRs are 15.9 dB and 26 dB. When $r = 1$, the corresponding SNRs are 15.2 dB and ∞. With the increase of r, the performance of outer-8PSK is getting better, while that of inner-2ASK is getting worse, but the performance of outer-8PSK is always better than that of inner-2ASK. This means that outer-8PSK is suitable for higher QoS signals, and inner-2ASK fits for lower QoS signals. When the channel condition is poor, increasing r will give priority to meeting the BER requirements of outer-8PSK, sacrificing the lower QoS service, so that the higher QoS service can be guaranteed. When the channel condition is good, we can reduce r to satisfy the BER requirements of outer-8PSK and inner-2ASK, thus all the QoS services can be guaranteed.

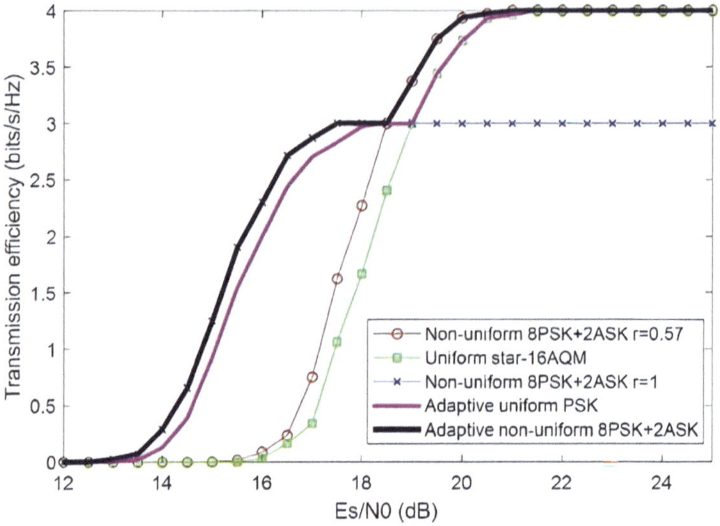

Fig. 2.4 Transmission efficiency

Figure 2.4 illustrates the transmission efficiency of several modulation schemes: adaptive non-uniform 8PSK + 2ASK modulation, adaptive uniform PSK modulation and uniform n-level ($n = 3, 4$) modulation. The results indicate that the adaptive non-uniform 8PSK + 2ASK modulation has better transmission efficiency than that of the adaptive uniform PSK modulation. Besides, the non-uniform 8PSK + 2ASK with $r = 0.57$ has the highest transmission efficiency for $SNR > 17.5dB$ while the non-uniform 8PSK + 2ASK with $r = 1$ achieves the best efficiency for SNR in 13–17.5 dB, which means that with the adjustment of r, the system efficiency can be improved efficiently.

2.2 Adaptive 3-Dimensional Hierarchical QAM

In traditional optical networks, a fixed modulation format is preferable for it has a simple structure with low cost. Nowadays internet applications are changing dramatically with the development of 5G because various information systems have different QoS requirements. Moreover, the channel conditions are complex in recent optical communication systems. Therefore, it is difficult for a single modulation format to achieve multi-service transmission. Multi-dimensional modulation formats have been used to improve the spectral efficiency, and the trellis coded modulation with three-dimension 9QAM mapping based on a novel probabilistic shaping method is proposed in Ref. [5]. The three-dimensional adaptive modulation and coding for transmission system is demonstrated in Refs. [6, 7]. Three-dimensional probabilistically shaped CAP (Carrierless Amplitude Modulation/Phase

Modulation) modulation based on constellation design using regular tetrahedron cells is proposed in Ref. [8], which demonstrated the proposed uniform 3D-CAP-16 outperforms the traditional counterparts by 1.1 dB receiver sensitivity gain. Moreover, Hierarchical modulation (HM) technology can be employed to adaptively adjust the spacing of constellation points according to channel conditions to maximize the efficiency of the system for meeting the QoS requirements.

In the literature, hierarchical quadrature amplitude modulation, phase-shift keying (PSK) and pulse amplitude modulation (PAM) modulation schemes have been proposed. The authors proposed an adaptive hierarchical PSK to transmit various traffic with different QoS requirements [4]. An adaptive hierarchical QAM scheme for enhanced bandwidth and power utilization is proposed in Ref. [9]. Hierarchical distribution matches for probabilistically shaped coded modulation, which can achieve performance close to the Shannon limit [10]. A non-uniform bit-mapping signal based on an InP transmitter module is demonstrated [11]. The authors proposed multi-level layered modulation in the spatial domains [12]. A new HM scheme was reported to increase the diversity gain through constellation rotation to improve the performance of low parity signals [13]. Hierarchical PAM4 is used in passive optical networks to redistribute power margin among users with different path losses, achieving an increase in the power budget by 3.5 dB [14]. In addition, the coded modulation schemes based on HM have been proposed [15, 16]. However, the mentioned schemes have used two-dimensional modulation without taking layered multi-dimensional modulation into consideration.

In this subsection, we propose a 3-Dimensional adaptive hierarchical QAM scheme (3D-AH-8/64QAM), which can improve the system power efficiency and reduce the bit error rate. The simulation results show that the proposed scheme can adapt to different channel condition and improve the system efficiency. Under the condition of the same spectral efficiency, the proposed 3D-AH-8/64QAM achieves maximum 3.3 dB and 5.7 dB SNR gains at the BER of HD-FEC threshold compared with 2D-8QAM and 2D-64QAM, respectively.

2.2.1 The Principle of 3D-AH-8/64QAM

The proposed 3D-AH-8/64QAM constellation is shown in Fig. 2.5a. A 3D-AH-8/64QAM signal contains 6 bits b6b5b4b3b2b1, in which the higher three bits b6b5b4 have higher priority (HP), and the lower three bits b3b2b1 have lower priority (LP). b6b5b4 using in 3D-8QAM modulation format is called 3D-HP-8QAM, which is represented by 8 blue cubes in Fig. 2.5a. The b3b2b1 used in 3D-8QAM modulation format is called 3D-LP-8QAM, which is represented by 8 black dots on the blue cube in Fig. 2.5a. Each blue cube and the points on it form a subset of 3D-AH-8/64QAM constellation, denoted as S_i

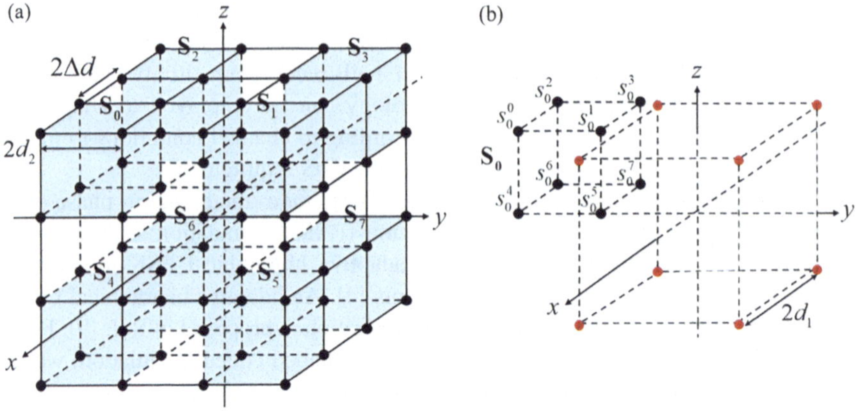

Fig. 2.5 The adaptive nonuniform modulation scheme based on three-dimensional space for 3D-AH-8/64QAM (**a**) the structure of 3D-AH-8/64QAM constellation presented with black dots on the blue cube. (**b**) The subset construction of 3D-AH-8/64QAM

$$S_i = \{s_i^0, s_i^1, \cdots, s_i^7\} i = 1, 2, \cdots, 8 \tag{2.6}$$

The 3D-AH-8/64QAM constellation can be expressed as a set of S_i

$$S = \{S_0, S_1, \cdots, S_7\} \tag{2.7}$$

It can be seen that the 3D-AH-8/64QAM constellation contains 64 signal points. The construction process of 3D-AH-8/64QAM is shown in Fig. 2.5b, the higher three bits b6b5b4 are mapped to a standard 3D-8QAM constellation in the shape of a regular hexahedron by a 3D mapper with a side length of $2d_1$, represented by a red dot in Fig. 2.5b. Each point of the uniform 3D-8QAM constellation is further divided into 8 points to form a small 3D-8QAM constellation with a side length of $2d_2$, which are represented by black dots in Fig. 2.5b. After the superposition of two layers of constellations, a 3D-AH-8/64QAM constellation subset is formed.

In Fig. 2.5a, $2\Delta d$ represents the minimum distance between adjacent S_i subsets, which is the minimum distance of the 3D-HP-8QAM constellation point. $2d_2$ represents the minimum distance between the constellation points of a S_i subset, which is the minimum distance of the 3D-LP-8QAM constellation points. Δd satisfies the following relationship between d_1 and d_2:

$$\Delta d = d_1 - d_2 \tag{2.8}$$

Priority parameters λ:

$$\lambda = \frac{d_2}{\Delta d} \tag{2.9}$$

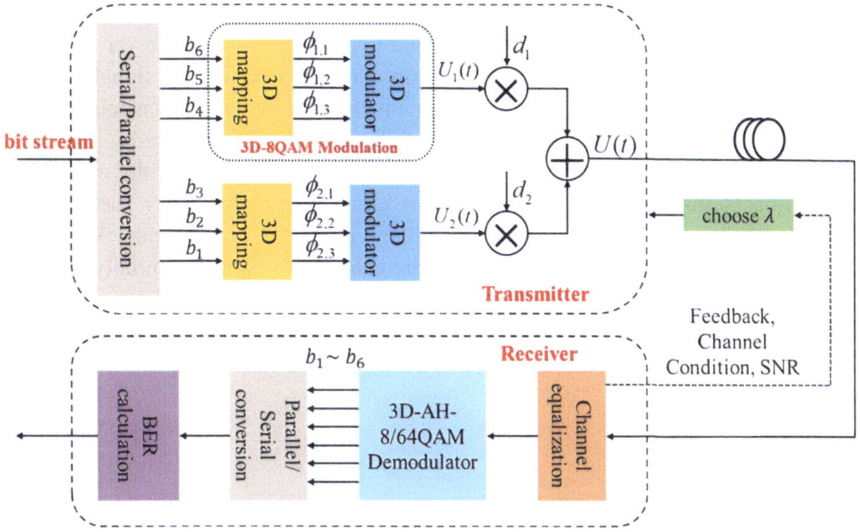

Fig. 2.6 The scheme of 3D-AH-8/64QAM system

According to $0 \leq \lambda \leq 1$, $0 \leq 2d_2 \leq d_1$, when $\lambda = 1$, the constellation is standard 3D-64QAM signal, and when $\lambda = 0$, the constellation becomes 3D-8QAM.

The block diagram of the 3D-AH-8/64QAM scheme system is shown in Fig. 2.6. At the transmitter, a set of random sequences is transformed into six groups of bitstreams b6b5b4b3b2b1 by serial-to-parallel conversion, where b6b5b4 form the signal $U_1(t)$ by a 3D-8QAM modulator. Similarly, b3b2b1 by 3D modulation is to generate the signal $U_2(t)$. The transmitted signal $U(t)$ at time t is expressed as a sum of $U_1(t)$ and $U_2(t)$:

$$U(t) = d_1 U_1(t) + d_2 U_2(t) \tag{2.10}$$

Where $U_1(t)$ and $U_2(t)$ is given by:

$$U_i(t) = \sum_{n=1}^{N} \varnothing_{i,n} \cdot \Phi_n(t), i = 1, 2 \tag{2.11}$$

In Eq. (2.10), N refers to the constellation dimension. In this scheme N = 3, $\varnothing_{i,n} \in \{-1, 1\}$. $\{\Phi_1(t), \Phi_2(t), \Phi_3(t)\}$ constitute a set of three-dimensional orthogonal basis functions. $\varnothing_{i,n} \in \{-1, 1\}$ means that each orthogonal function is modulated with binary amplitude keying (2ASK), with reduced complexity compared to conventional 3D-64QAM. For conventional 3D-64QAM, $\varnothing_{i,n} \in \{-3, -1, 1, 3\}$. It is necessary to perform 4ASK modulation on the orthogonal basis function, thus the complexity of its three-dimensional modulation and demodulation is greatly increased.

After the signal $S(t)$ passes through the Gaussian white noise channel, the HP bits and LP bits of the signal will be demodulated respectively at the receiver, and then the BER is calculated. In the conventional soft decision, the received signal needs to be compared with 64 constellation points in the 3D-8/64QAM constellation. In contrast, the layered demodulation method only needs to be compared with 8 constellation points in every two stages, which greatly reduces the difficulty of demodulation decisions. Moreover, the receiver continuously monitors the channel condition and feeds it back to the transmitter via the feedback signal. When the channel condition is poor, we can increase the value of λ thus preferentially transmit 3D-HP-8QAM at the cost of reducing the BER performance of 3D-LP-8QAM or even only transfer 3D-HP-8QAM signals. In good channel condition, all points can be assigned with bitstreams and the BER performance of 3D-LP-8QAM will also be improved.

2.2.2 Simulation and Results

The constellation diagram under different λ is shown in Fig. 2.7. Figure 2.7 a–c show the received constellation of 3D-AH-8/64QAM signals when λ is 1, 0.3, and 0 respectively. It can be seen from Fig. 2.7, the distance between adjacent 3D-HP-8QAM constellation points increases as λ decreases, while the sensitivity of the 3D-HP-8QAM signal will be improved and the bit error rate decreases. When $\lambda = 0$, the 3D-AH-8/64QAM constellation becomes a standard 3D-8QAM constellation. On contrary, the 3D-LP-8QAM constellation point spacing becomes larger when λ increases, consequently the sensitivity of 3D-HP-8QAM signal will be improved and the bit error rate decreases. When $\lambda = 1$, the 3D-AH-8/64QAM signal becomes a standard 3D-64QAM constellation with even spacing.

Figure 2.8 shows the performance of 3D-AH-8/64QAM signal compared with the 2D signal with the same spectral efficiency. From left to right, the BER curves are 3D-HP-8QAM at $\lambda = 0$ (uniform 3D-8QAM), 2D-8QAM, 3D-AH-8/64QAM at $\lambda = 1$ (uniform 3D-64QAM) and 2D-64QAM respectively. The 2D-8QAM constellation is shown on the right corner of Fig. 2.8. At the threshold of the BER is 1e-3, the SNR required for 3D-HP-8QAM and 2D-8QAM with spectral efficiency of

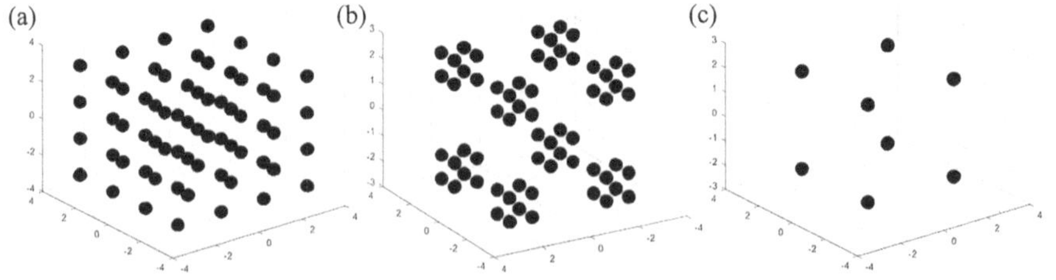

Fig. 2.7 3D-AH-8/64QAM constellation (**a**) $\lambda = 1$, (**b**) $\lambda = 0.3$, (**c**) $\lambda = 0$

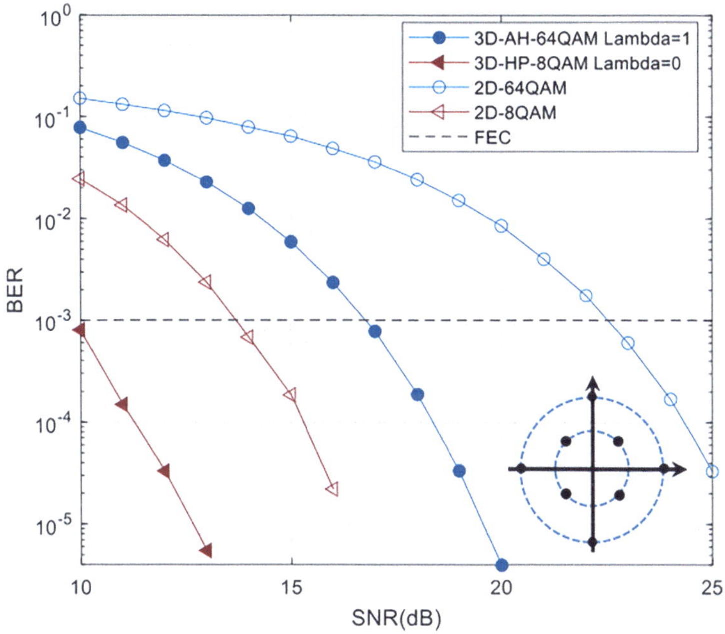

Fig. 2.8 BER performance of the 3D-AH-8/64QAM signal ($\lambda = 0.1$)

3 bit/s/Hz is 10 dB and 13.3 dB respectively. The SNR required for 3D-AH-8/64QAM and 2D-64QAM with 6 bit/s/Hz spectral efficiency is 16.8 dB and 22.5 dB respectively. According to the simulation results, the SNR gains obtained by 3D-8QAM and 3D-64QAM are 3.3 dB and 5.7 dB respectively, and the performance is improved apparently.

Figure 2.9 shows the performance of the 3D-AH-8/64QAM scheme under four cases where λ is ~$\{0.6, 0.7, 0.8, and 0.9\}$. The BER curves of 3D-HP-8QAM and 3D-LP-8QAM are distributed on both sides of the uniform 3D-64QAM. On the condition of $\lambda = \{0.9, 0.8, 0.7, 0.6\}$ and $BER = 1e - 3$, the required SNR for 3D-HP-8QAM and 3D-LP-8QAM are 15.5 dB, 17.1 dB; 15.0 dB, 17.6 dB; 14.2 dB, 18.0 dB; and 14.0 dB, 19.0 dB respectively. With the decreasing of λ, the performance of 3D-LP-8QAM becomes worse and that of 3D-HP-8QAM gets better. The receiver sensitivity of 3D-HP-8QAM is easier to improve than that of 3D-LP-8QAM with the change of λ. Therefore, the 3D-HP-8QAM signal is suitable for transmitting information with higher QoS requirements, and the 3D-LP-8QAM signal is suitable for transmitting information with lower QoS requirements.

Figure 2.10 shows the performance of 3D-AH-8/64QAM under different SNR ($BER < 1e - 3$). Figure 2.10a displays the minimum SNR threshold required for switching different modulation formats in the 3D-AH-8/64QAM scheme. According to the results in Fig. 2.10a, when the SNR is less than 16.6 dB, the 3D-HP-8QAM can be used for transmission. When the SNR is within the range of 16.6~17 dB, the uniform 3D-64QAM with the spectral efficiency of 6 bit/s/Hz can be used. The

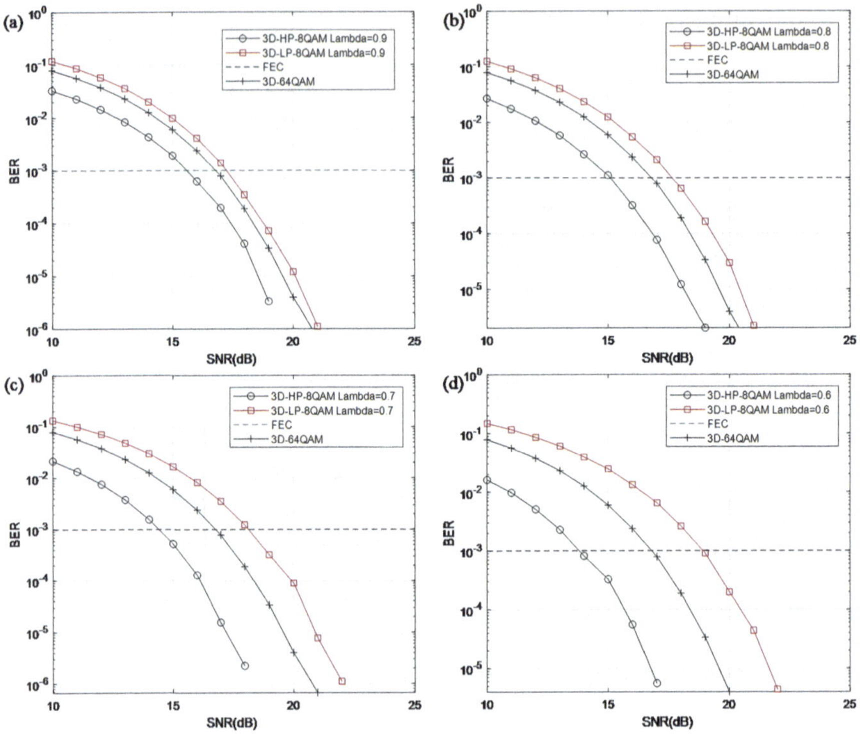

Fig. 2.9 BER performance of the 3D-AH-8/64QAM signal. (**a**) $\lambda = 0.9$, (**b**) $\lambda = 0.8$, (**c**) $\lambda = 0.7$, (**d**) $\lambda = 0.6$

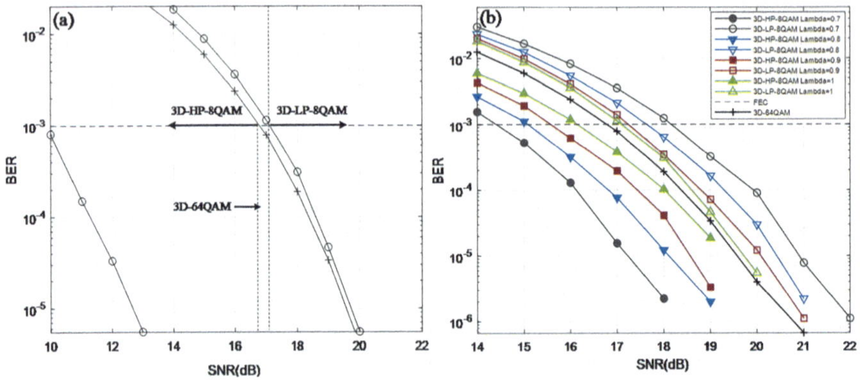

Fig. 2.10 The relationship between BER and SNR of 3D-AH-8/64QAM. (**a**) $\lambda = 0.1$, (**b**) $\lambda = 0.7$, 0.8, 0.9, 1

3D-LP-8QAM can be used when the SNR is greater than 17 dB. The channel condition changes can be realized by adjusting the value of λ to meet the system requirements. According to the different SNR thresholds, the different QoS requirements can be carried on the same signal to improve the system efficiency without changing any hardware [17].

Figure 2.10b shows the performance of the 3D-HP-8QAM signal and the 3D-LP-8QAM signal. In the proposed 3D-AH-8/64QAM modulation scheme, the services with lower QoS requirements can be sacrificed within an acceptable range by reducing λ, so that the services with higher QoS requirement can be guaranteed. For example, when the SNR is 18 dB, the BER of HP bits decreases and that of LP bits increases by adjusting λ from 1 to 0.9. And when the λ is adjusted to 0.8, the BER of HP bits can be further reduced, and the BER of LP bits increases. For the QoS requirements, λ can be appropriately increased to make the performance of 3D-HP-8QAM and 3D-LP-8QAM more balanced and improve the overall signal quality of service.

2.3 Probabilistic Shaping Four-Dimensional Modulation

With the rapid development of mobile communication networks, live video broadcast, telemedicine, and other communication services, the internet traffic is growing rapidly for center interconnection. Consequently, the development of large-capacity optical communication is necessary. The technology of ultra-high order modulation (UHM) is used to increase the spectral efficiency (SE) toward the Shannon limit and meet the growing demand for high transmission capacity. The first 1024 QAM single-carrier coherent optical transmission is demonstrated in, in which a 60 Gbit/s polarization-multiplexed signal at a SE of 13.8 bit/s/Hz was transmitted over 150 km. Similarly, probabilistically shaped 4096-QAM has been transmitted over 50.9 km at a potential SE of 15.9 bits/s/Hz and a net bit rate of 484.4 Gb/s. So far, the highest-order QAM constellations that have been demonstrated in optical systems are 10-GBd probabilistically shaped square 16384-QAM at a net bit rate of 223.8 Gb/s. However, as the modulation format order increases, the minimum Euclidean distance decreases, and thus the ability to resist noise declines. Therefore, it is critical to propose a novel modulation format to expand the Euclidean distance of constellation points and optimize the DSP algorithm to improve the system performance.

Therefore, the four-dimensional (4D) modulation format is an advanced optical modulation format that is generated by drawing into the check bit through modulation coding in the two quadrature-phase components and two polarizations of the optical field. It has attracted a lot of attention in recent years because it can increase the minimum Euclidean distance between symbols. The 4D modulation format uses Ungerboeck's set partitioning strategy to generate the parity by XOR operation of information bits, such as 128-level SP quadrature amplitude modulation (128-SP-QAM) and 2048-SP-QAM. Moreover, probabilistic shaping (PS) is a technology that can increase the mutual information of signals toward the Shannon limit and provide a flexible platform for rate adaptation. PS-based multi-dimensional signaling schemes have recently been reported to provide higher fiber nonlinear tolerance. The PS scheme, however, is incompatible with the SP-QAM modulation format due to the presence of parity bits.

In this section, the scheme of 4D PS-4096-QAM is proposed. The amplitude translation method is used to generate the parity for the SP modulation format.

2.3.1 System Model

Figure 2.11 illustrates the model of the proposed four-dimensional modulated SCHD system based on AT. In the transmitter, the information bits derived from a discrete memoryless binary source are fed to the probabilistic amplitude shaping (PAS) block, which produces symbols following the Maxwell-Boltzmann (MB) distribution for the 4 paths. The MB distribution can be expressed as

$$P_X(x) = \frac{e^{-\lambda x^2}}{\sum_{x' \in X} e^{-\lambda x'^2}}. \tag{2.12}$$

where x represents the $\sqrt{M}$-PAM constellation points and λ denotes the shaping factor. In a one-path PAS, the input uniformly distributed binary bit sequence is divided into a bit sequence b^k of length k and a bit sequence b^j of length j. The sequence b^k is fed into the distribution matcher (DM) to produce non-uniformly distributed positive amplitude $\sqrt{M}/2-\text{PAM}$ symbols S_+^n, with the alphabet

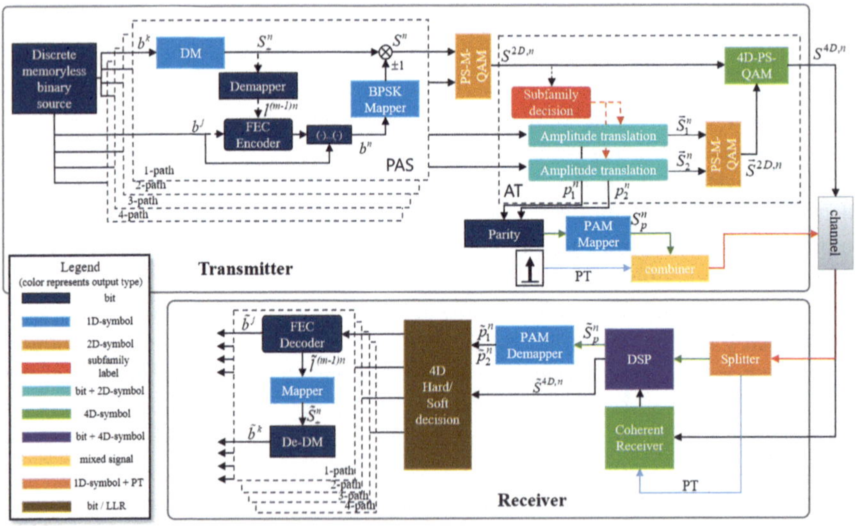

Fig. 2.11 Modeling of SHCD system based on 4D modulation

$A = \{1, 3, \ldots, \sqrt{M}\}$. The symbols S_+^n from the DM are then reflected in a PAM demapper as binary reflected Gray code (BRGC) sequence $l^{(m-1)n}$, which are combined with a sequence of bits b^j to be fed into the forward error correction (FEC) encoder to produce parity bits. The value of m is $\log_2 \sqrt{M}$. These parity bits and bit sequence b^j form a new bit sequence b^n. The bit sequence b^n is converted by the binary phase shift keying (BPSK) mapper from bit to n amplitudes, and i.e., 0 is mapped onto -1 and 1 is mapped onto 1. These ± 1 act as signs for S_+^n to form $\sqrt{M} -$ PAM signals S^n.

After PAS, the S^n of path 1 and path 2 form a sequence of 2D PS-M-QAM symbols $S^{2D,\, n}$, which can be interpreted as signals in the X polarization on the light field. The symbols of path 3 and path 4 are to be determined based on the subsets and subfamilies satisfied for the 4D symbols according to the sequence $S^{2D,\, n}$, and the symbols of the corresponding paths are to be amplitude translated for the formation of valid 4D modulation signals. Following this process, the translated symbols $\vec{S}_1^n$ and $\vec{S}_2^n$ and the parity bits p_1^n and p_2^n are produced. The 2D PS-M-QAM symbol sequence $\vec{S}^{2D,n}$, formed with $\vec{S}_1^n$ and $\vec{S}_2^n$, is regarded as a signal in the Y polarization. Eventually $S^{2D,\, n}$ and $\vec{S}^{2D,n}$ are formed to be a complete 4D signal sequence $S^{4D,\, n}$. On the other hand, p_1^n and $p_2^n$are mapped onto a sequence of PAM-4 signals, which are combined with an optical pilot tone (PT) to form a mutually non-interfering mixed signal and transmitted with $S^{4D,\, n}$ over a fiber optic channel.

At the receiver, the transmitted 4D symbols are received via a coherent receiver, while the mixed signal is split by a splitter into two branches. The PT branch is employed as the local oscillator (LO) of the coherent receiver, and the PAM signal branch and coherently received 4D symbols are ideally eliminated from other impairments by digital signal processing (DSP), given the fact that the signals are only impaired by Gaussian noise. The PAM signal $\tilde{S}_p^n$ is utilized by a PAM demapper to obtain the parity bits $\tilde{p}_1^n$ and $\tilde{p}_2^n$ for the hard or soft decision of the recovered 4D signal $S^{4D,\, n}$. The decided 4D signal is divided into four paths for FEC decoding, mapping and de-DM respectively, to finally obtain the transmitted binary bits $\tilde{b}^k$ and $\tilde{b}^j$. More details of the AT based SP-4D modulation and demodulation are described in this section and a preliminary version can be found in [18].

The four-dimensional modulation based on amplitude translation keeps the characteristics of the probability distribution of the signal constellation while expanding the MED of the signal constellation with the set partitioning principle to enhance the noise tolerance. The 1D nonuniform PAM-4 signal constellation produced by the PS is divided into two 1D subsets, C_{1D}^1 and C_{1D}^2, as shown with the 16-QAM-based constellation in Fig. 2.12. Then, the two 1D subsets of the I- and Q-components are combined pairwise into four 2D subsets A B C and D, where $A = C_{1D}^1 \cup C_{1D}^1, B = C_{1D}^1 \cup C_{1D}^2, and\ C = C_{1D}^2 \cup C_{1D}^1, D = C_{1D}^2 \cup C_{1D}^2$. The two 2D subsets of X -polarization and Y-polarization are subsequently combined pairwise into four 4D subfamilies A, B, C, and D, which are called subfamily constraints (SFC). The MED of the 4D constellations after set partitioning is expanded to twice

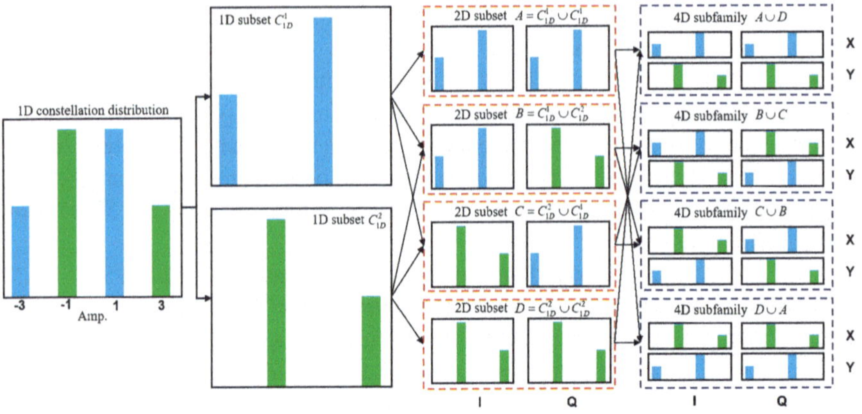

Fig. 2.12 Schematic of the partitioning of non-uniform constellation sets from 1D to 4D

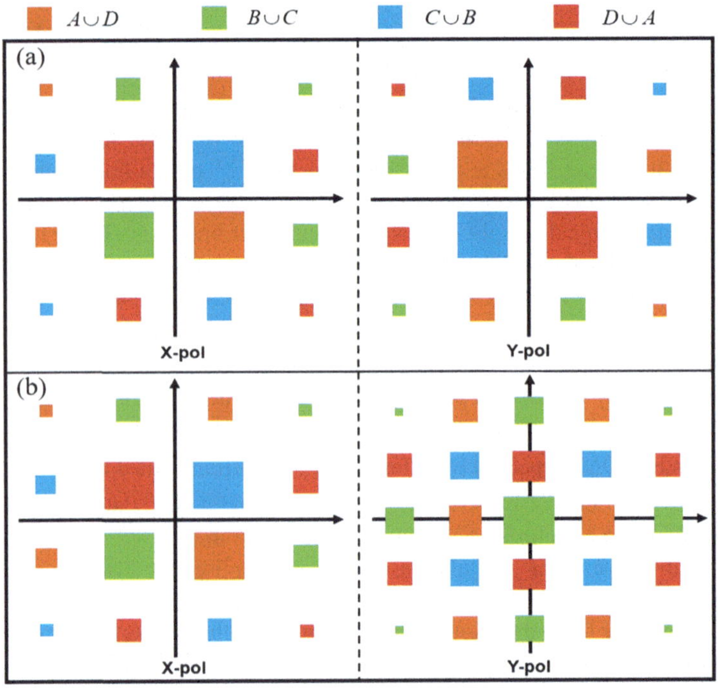

Fig. 2.13 Projections of 4D constellations on X and Y polarizations before (**a**) and after (**b**) AT

that of the original one. The projections of the nonuniform 4D constellations in X-polarization and Y-polarization are shown in Fig. 2.13a, with the size of the squares representing the probability magnitude of the constellation points.

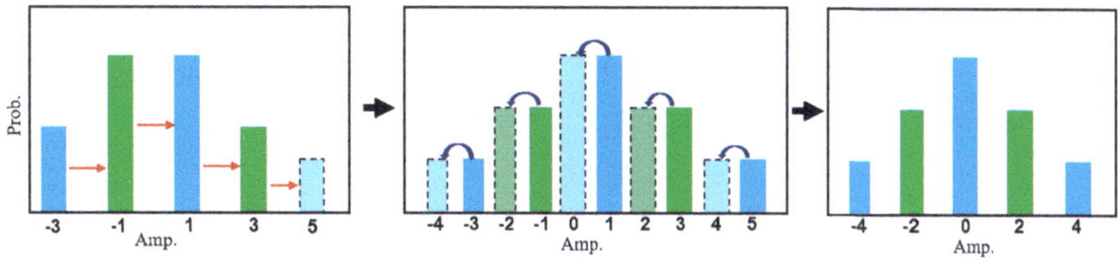

Fig. 2.14 Schematic diagram of amplitude translation and signal constellation variation

For the 4D signal to satisfy the above 4D subfamily partitioning rules, the AT is required. Following the PAS of the four paths, the 1D symbols of path 1 and path 2 first form the 2D signals of the X-polarization carrying

$$S_x^{2D} = S_1 + iS_2, \tag{2.13}$$

where S_1 and S_2 are the symbols of path 1 and path 2 respectively, in a symbol period. The corresponding 4D subfamily is chosen according to C. The 1D subset of symbols in path 3 and path 4 that is required to satisfy this subfamily is decided. A discussion is then divided into two scenarios.

(1) If a 1D symbol on a given path mismatches the current SFC, the 1D symbol is translated to a neighboring signal point (forward translation is employed in this section), as shown by the red arrow in Fig. 2.14, i.e., a complementary 1D subset, and then the binary label is set to 1. During the AT, the end points are translated beyond the limits of the original constellation, adding a new amplitude, e.g., in Fig. 2.14, the translation of a PAM-4 symbol of amplitude 3 produces a symbol of amplitude 5. The equation for the example of path 3 is described as

$$\underline{S}_1 = S_3 + d : label \rightarrow 1, \tag{2.14}$$

where d represents the MED in the 1D constellation.

(2) If the one-dimensional signal is consistent with the SFC, no translation is required and the labeling bit is 0, which can be described as

$$\underline{S}_1 = S_3 : label \rightarrow 0. \tag{2.15}$$

During the AT, a direct current (DC) is induced, leading to the constellations following a quasi mb distribution, which impacts the power efficiency and performance of the system adversely. It is assumed that the 1D constellation of signals before AT features an alphabet $X = \{\pm 1, \pm 3, \ldots, \pm \sqrt{M} - 1\}$ following the MB distribution with parameter λ. All alphabets in this section are sorted in ascending order of element size. The set $L = \{0, 1\}$ of labeled bits produced by the 4D signal follows a (0-1) distribution with parameter 1/2. X and L are independent of each other. The alphabet of signals after AT is $Y = \{\pm 1, \pm 3, \ldots, \pm\sqrt{M} - 1, \sqrt{M} + 1\}$. For a label of 0, the distribution of the signal constellations after AT is

$$P_r(y_i) = \frac{e^{-\lambda y_i^2}}{\sum_{x' \in X} e^{-\lambda x^2}}. \tag{2.16}$$

For the label of 1, the amplitude of the signal point is translated with a probability of

$$P_{\mathrm{y}}(y_i) = \begin{cases} 0 & i = 1 \\ P_{\mathrm{Y}}(y_{\mathrm{i+1}}) & 2 < i \leq \sqrt{M} + 1 \end{cases}. \tag{2.17}$$

The probability mass function (PMF) of Y is:

$$\begin{aligned} P_Y(y_i) &= P\{Y = y_i\} \\ &= P\{Y = y_i, L = 0\} + P\{Y = y_{\mathrm{i}}, L = 1\} \\ &= \begin{cases} \dfrac{e^{-\lambda y_i^2}}{2\sum_{x' \in X} e^{-\lambda x'}} & i = 1 or i = \sqrt{M} + 1 \\ \dfrac{e^{-\lambda y_i^2} + e^{-\lambda y_{i-1}^2}}{2\sum_{x' \in X} e^{-\lambda x'}} & 1 < i \leq \sqrt{M} \end{cases} \end{aligned} \tag{2.18}$$

The expectation of Y is 1, which means that there is a DC of magnitude 1 for signals with an alphabet of Y. In [18], the points at the end are then translated to the head, forming a ring structure to slightly mitigate the DC, but the adverse effect is still present. In this section, we optimize the translated signal constellation by de-DCing the amplitude of the translated signal, as shown in Fig. 2.14. The PAM constellation with a MED of d and a value of DC of d/2 can be derived even further based on the above analysis. Therefore, the nonuniform $(\sqrt{M} + 1) -$ PAM constellation obtained from AT and de-DC features an alphabet $Y' = \{0, \pm 2, \pm 4, \ldots, \pm\sqrt{M}\}$ and the PMF of Y' is

$$P_{r_i}(y_i') = \begin{cases} \dfrac{e^{-\lambda y_i^2}}{2\sum_{x' \in X} e^{-\lambda x^2}} & i = 1 \quad or \quad i = \sqrt{M} + 1 \\ \dfrac{e^{-\lambda y_i^2} + e^{-\lambda y_{i+1}^2}}{2\sum_{x' \in X} e^{-\lambda x^2}} & 1 < i \leq \sqrt{M} \end{cases} \tag{2.19}$$

And Eqs. (2.13) and (2.14) can be updated as

$$\vec{S}_1 \& = \underline{\vec{S}}_1 - \frac{d}{2} = S_3 + \frac{d}{2} : label \rightarrow 1, \tag{2.20}$$

$$\vec{S}_1 \& = \underline{S}_1 - \frac{d}{2} = S_3 - \frac{d}{2} : label \rightarrow 0, \tag{2.21}$$

The symbols $\vec{S}_1$ and $\vec{S}_2$ of paths 3 and 4 after translation and de-DC form the 2D symbols of the Y-polarization carrying as

$$\vec{S}_y^{2D} = \vec{S}_1 + i\vec{S}_2. \tag{2.22}$$

Symbols S_x^{2D} and $\vec{S}_y^{2D}$ eventually form a 4D symbol S^{4D}. The projection of S^{4D}on the X-polarization and Y-polarization is shown in Fig. 2.13b.

The 4D signal demodulation can be realized with hard decision (HD). Based on the hard decisions of the 4D symbols in accordance with the 4D Euclidean distance, the signal in the Y polarization is translated into a signal in the backward direction. For the label bit of 1, the amplitude of the corresponding signal component is translated backward. For binary label bits of 0, no translation is required. After completing the translation of all symbols, the decoding of the PAS is performed. In this section we present new 4D soft decision (SD) based bit-wise decoding for 4D-PS-QAM. It is considered in the memoryless 4D Gaussian channel Y = X + Z, where Z = [Z1, Z2, Z3, Z4] is a 4-dimensional zero-mean Gaussian real random variable. The total variance of the noise is σ_z^2. The log-likelihood ratio (LLR) for each bit-level i calculation equation is described as [19]

$$L_i = \log \frac{\sum_{x \in X_i^1} \exp\left(-\frac{\| y - x \|^2}{\frac{\sigma_z^2}{4}}\right)}{\sum_{x \in X_i^0} \exp\left(-\frac{\| y - x \|^2}{\frac{\sigma_z^2}{4}}\right)} \approx \frac{\sigma_z^2}{4}\left(\min_{\mathbf{x} \in \mathbf{X}_{4,j}^0} \| \mathbf{y} - \mathbf{x} \|^2 - \min_{\mathbf{x} \in \mathbf{X}_{4,j}^1} \| \mathbf{y} - \mathbf{x} \|^2\right) \tag{2.23}$$

where $i = 1, 2, \ldots, 2\log_2(M).X_{4,i}^b$ is the alphabet of 4D constellation symbols labeled by bit b ∈ {0, 1} at the i-th bit. $\frac{\sigma_z^2}{4}$ represents the variance of each real-dimensional Gaussian channel represents the variance of each real-dimensional Gaussian channel $\|y - x\|^2 = |y_1 - x_1|^2 + |y_2 - x_2|^2 + \cdots + |y_4 - x_4|^2$. Defining a set $F = \{A \cup D, B \cup C, C \cup B, D \cup A\}$ consisting of SFCs, it is possible to rewrite the two terms on the right side of the Eq. (2.23) as

$$\min_{x\in X^b_{4,j}} \| y-x \|^2 = \min_{SFC\in F}\left(\min_{x\in X^b_{4,SFC,j}} \| y-x \|^2\right). \tag{2.24}$$

where $X^b_{4,SFC,i}$ represents the alphabet of 4D constellation points that satisfy the current SFC at the $i-th$ bit. We define the symbol dimension associated with the $i-th$ bit as bit-related dimension (BRD), which gives

$$\begin{aligned}\min_{x\in X^b_{4,SFC,i}} y-x^2 &= \min_{(x_1,\ldots,x_4)\in X^b_{4,SFC}}\left(\sum_{m=1}^{4}(y_m-x_m)^2\right)\\ &= \sum_{\substack{m=1\\ m\neq BRD}}^{4}\min_{x_m\in X^b_{1,SFC-m}}(y_m-x_m)^2+Q^b\end{aligned} \tag{2.25}$$

where $X^b_{1,SFC-m,i}$ is the alphabet of 1D constellation that satisfy the dimension associated with the $i-th$ bit of the current SFC. Q^b is jointly determined by BRD and parity produced by AT. For (1) BRD is an untranslated dimension, i.e., two real signal dimensions on the X polarization, or (2) BRD is a translated dimension, i.e., two real signal dimensions on the Y polarization, and the corresponding parity is 0. The symbols of this dimension are not translated, then there is

$$Q^b = \min_{x_{BRD}\in X^b_{1,SFC-BRDj}}(y_{BRD}-x_{BRD})^2. \tag{2.26}$$

For (3) BRD is a translation dimension and the corresponding parity is 1. The symbols of this dimension are translated, then, the

$$Q^b = \min_{x_{BRD}\in \vec{X}^b_{1,SFC-BRD,i}}(y_{BRD}-x_{BRD})^2. \tag{2.27}$$

Where $\vec{X}^b_{1,SFC-BRD,i}$ is a subset of $X_{1,\ SFC\ -\ BRD,\ i}$ not containing $\min_{x_{BRD}\in X^b_{1,SFC-BRD,i}}(y_{BRD}-x_{BRD})^2$. Meanwhile, the corresponding required variation is $Q^{1\ -\ b}$, as

$$Q^{1-b}\& = \min_{x_{BRD}\in X^{1,b}_{1,SFC-BRD}}(y_{BRD}-x_{BRD})^2 \tag{2.28}$$

$$\vec{\mathbf{X}}^{1-b}_{1,SFC-BRD,i}\& = \mathbf{X}^b_{1,SFC-BRD,i}\cup\left\{\min_{x_{BRD}\in X^b_{1,SFC-BRD,i}}(y_{BRD}-x_{BRD})^2\right\} \tag{2.29}$$

In summary, the LLR for each bit-level i of the 4D constellation is calculated as

$$
\begin{aligned}
L_i \approx \frac{\sigma_z^2}{4} \Bigg\{ & \min_{SFC \in F} \left(\sum_{\substack{m=1 \\ m \neq BRD}}^{4} \min_{x_m \in X^0_{1,SFC-mJ}} (y_m - x_m)^2 + Q^0 \right) \\
& - \min_{SFC \in F} \left(\sum_{\substack{m=1 \\ m \neq BRD}}^{4} \min_{x_m \in X^1_{i,SFC-m,j}} (y_m - x_m)^2 + Q^1 \right) \Bigg\}
\end{aligned}
\tag{2.30}
$$

where Q^0 and Q^1 are given by Eqs. (2.26)–(2.29).

2.3.2 Simulation Results

The performance of the translated constellations is evaluated by comparing the power efficiency of the 2D signal with that of the 4D signal, basing the constellation in PS-4096-QAM. We conventionally refer to the ordinary dual-polarization signals as 2D signals in both the simulation and experimental results that follow. The asymptotic power efficiency for M-ary constellation is described as

$$
\gamma = \frac{d_{min}^2 H}{4E_s} \tag{2.31}
$$

where d_{min}^2 represents the square of MED and E_s is the average symbol energy. Figure 2.15 illustrates the asymptotic power efficiency of the three modulation schemes for different constellation entropies H. For high entropy constellations (H > ~13 bit/symbol), the gain of γ for the signal is ~6 dB for both 4D schemes. For low entropy constellations, the gain of the constellations described in [18] decreases to ~ 5 dB with an entropy H of 8.1 bit/symbol, which makes the modulation scheme more suitable for constellations with high entropy. In this section, the constellations are optimized to mitigate the DC penalty for performance gain, with a gain of about 5.5 dB, which increases the applicability of low entropy constellations.

Subsequently, we performed simulations of the 4D-PS-4096-QAM signal in an additive white Gaussian noise (AWGN) channel. To investigate the performance of the optimized 4D-PS signal with HD and SD, generalized mutual information (GMI) based on both HD and SD decoder is evaluated for AIR. Figure 2.16 illustrates the AIR as a function of SNR for 2D-PS-4096-QAM with HD, 4D-PS-4096-QAM with HD and SD in seven entropy constellations. AIR AIR_{HD} for HD and AIR_{SD} for SD are calculated separately as [20, 21],

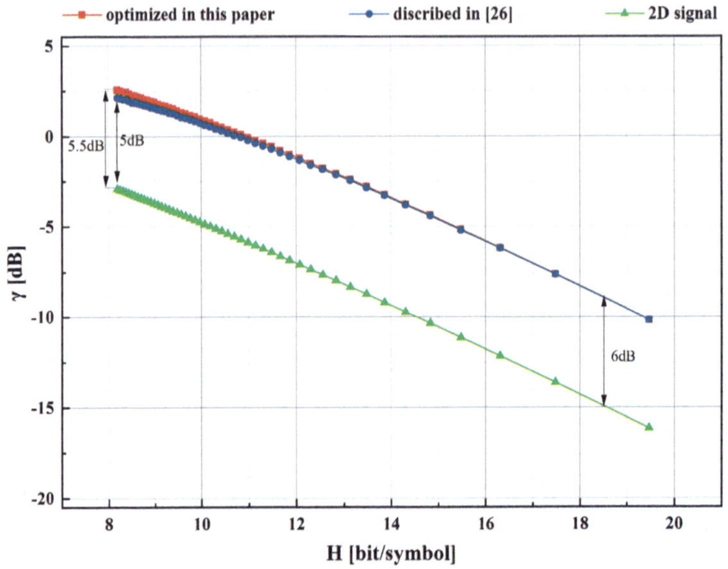

Fig. 2.15 Asymptotic power efficiency of three modulation schemes

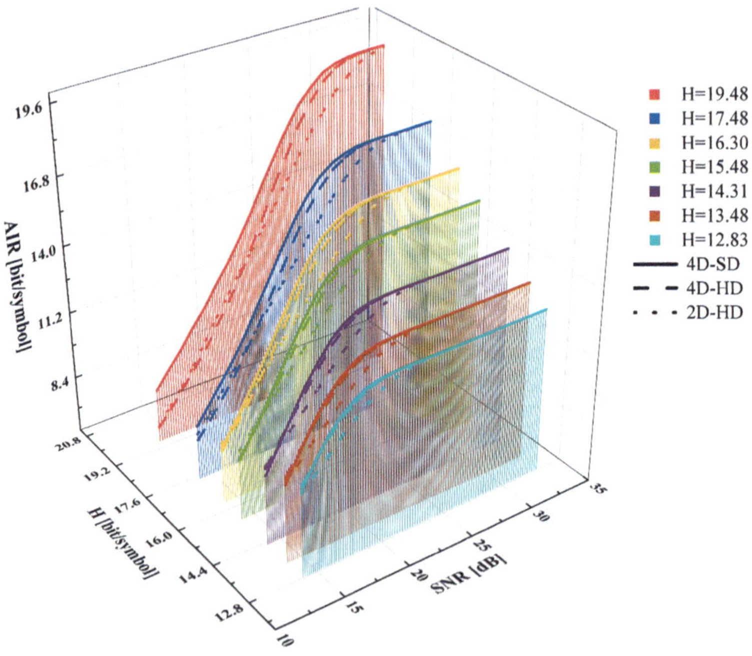

Fig. 2.16 AIR as a function of SNR with Gaussian channels for different entropy constellations

$$AIR_{HD} = GMI_{HD} = H[1 - (-\varepsilon\log_2\varepsilon - (1-\varepsilon)\log_2(1-\varepsilon))]$$
$$AIR_{sD} = GMI_{sD} \approx H - \frac{1}{D}\min_{s\geq 0}\sum_{k=1}^{m}\sum_{l=1}^{D}\log_2\left(1 + e^{s(-1)^{q_{k,j}}L_{q,j}}\right) \tag{2.32}$$

where ε is error probability, D represents the number of symbols, and c is the binary information sequence. Approximating the minimization of s, the GMI in s is concave. It can be seen that for all values of H, there is gain in 4D-SD with respect to both 4D-HD and 2D-HD, but the gap is slightly wider and the gain is more pronounced for higher entropy values. It is worth noting that for low SNR, there is an intersection between the curves of 4D-HD and 2D-HD, which implies that the gain of 4D modulation exists in the middle and high SNR regions.

2.4 ATTCM Concatenated Coded Modulation

At present, the demands of the optical fiber communication are growing dramatically and rapidly. Optical access network is a strong candidate for connecting the backbone network with the end users, which provides high speed and high reliability for bandwidth hungry applications [20, 21]. Currently commercial access systems mainly apply lean single carrier. However, the increasing demands of users for extended bandwidth resources are pushed forward. As a result, the multicarrier technology has been proposed in the past few years, the fourth generation (4G) cellular systems based on OFDM technology have been optimized to provide high data rates and reliable coverage to mobile users [22]. Nevertheless, OFDM is susceptible to inter-carrier interference and synchronization errors. Additional overhead such as cyclic prefix (CP) is needed to alleviate the interference [22]. While generalized frequency division multiplexing (GFDM) based on multicarrier filter bank is a strong modulation technique for the fifth generation (5G) standard [23–26]. The main characteristic of GFDM is low out of band emission, insensitive to frequency offset and phase noise. Cyclic prefix is added to each GFDM block instead of each GFDM symbol. The peak-to-average power ratio (PAPR) can be reduced by designing the flexible pulse shaping of individual sub-carriers [25, 26].

In addition, GFDM technology with high spectral efficiency enables the access system to support a huge number of optical network units. They are more susceptible to transmission impairments from channel and components in optical access network, which leads to inter-symbol interference (ISI) and performance degradation of the system [25]. Therefore, the forward error correction (FEC) codes can be employed to improve the BER performance by providing extra gain to the optical access network; it is a powerful technique in increasing the system's margin [27–29]. Turbo code as one of the FEC code, has very good performance for error correction. It not only has superior performance in high noisy environment, but also has strong anti-fading and anti-interference ability. Moreover, the ability to provide communication near the Shannon limit in various harsh conditions has been demonstrated. Turbo code has been as deep space communication standard by the United States advisory committee of space data system which provides an important background for the research of Turbo code implementation [30]. However, Turbo code has the problem of error floor, this phenomenon will seriously affect the high reliability in the actual communication. The technology of combing turbo code with trellis coded modulation (TCM) can improve spectrum efficiency and has robust error correction quality with simple construction where the error floor problem can be avoided, which makes it an attractive solution for optical access network in the near future [31–35].

In this subsection, we propose and experimentally demonstrate a novel concatenated coded modulation (adaptive turbo trellis-coded modulation (ATTCM)) in 4 Gbit/s GFDM optical access system. The code weight decision algorithm (CWD) is used for adaptive code. The decoder of ATTCM used Log-MAP algorithm and reduced the complexity of the receiver. With these techniques combined, the proposed scheme provides significant performance improvement for access system.

2.4.1 *Principle of System Model*

The schematic diagram of concatenated coding modulation GFDM is illustrated in Fig. 2.17. It is composed of ATTCM module and GFDM modulation. The principle of each part is described in the following section.

2.4.1.1 ATTCM Encoder and Decoder

In the section of ATTCM encoding, we firstly use code weight decision algorithm (CWD) before injecting the input data into the encoder as shown in Fig. 2.18. The code weight decision algorithm is used for improving the code weight of the input data bits. Let the input data pass the encoder directly if the weight of the code is higher than a certain value of its length ($W > L/n$, W present the code weight, L is the length of the input data, n is positive integer), otherwise, all bits of the input data will be inverted and then the inverted bits will be marked. The encoder implementation sends three sub-blocks of bits. The first sub-block is a complete set after fulfilling the condition $W > L/n$. The second and the third sub-blocks are encoded respectively

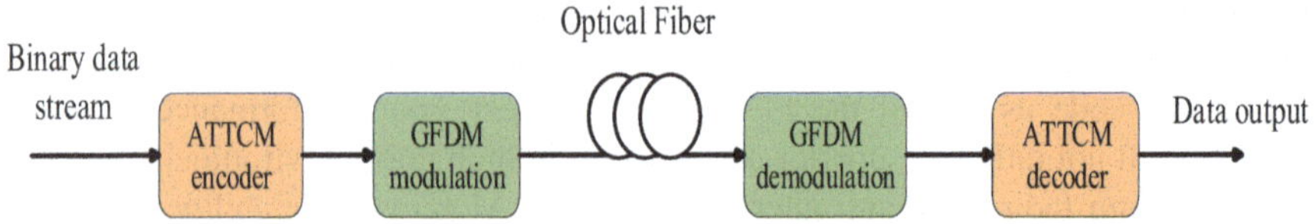

Fig. 2.17 Schematic diagram of concatenated coded GFDM signal

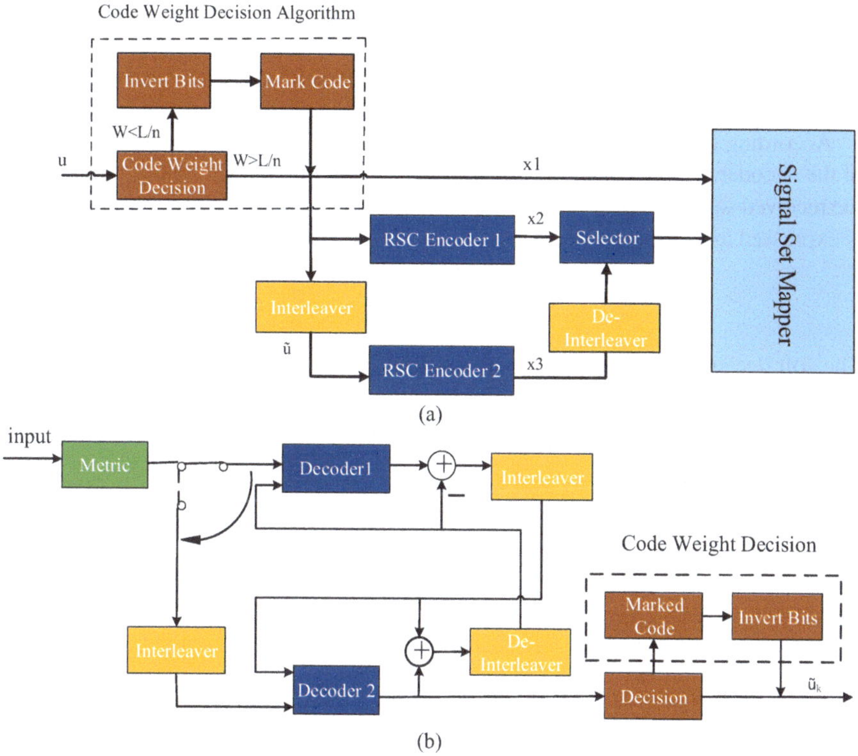

Fig. 2.18 The scheme of ATTCM with CWD algorithm (**a**) ATTCM encoder (**b**) ATTCM decoder

where the third part is sent to an inter-leaver. These two redundant but different sub-blocks are encoded by the same RSC (Recursive Systematic Convolution) encoder. Since the encoder is systematic, the output of the upper encoder and the output of the de-interleaver are different only in the parity bits, while the remaining bits are identical. A selector is then used to alternatively switch between the parity bits from the upper and lower encoder. Finally, two combined sub-blocks are then mapped onto M-ary symbol.

Figure 2.18b shows the ATTCM decoding section. At the beginning, the symbol log-likelihood ratios (LLRs) are calculated and the corresponding soft metric is provided to the decoder. For the first decoding stage, the soft metrics corresponding to the odd symbols (t is odd) are fed to the decoder1, the symbol-by-symbol maximum a-posteriori (MAP) algorithm is employed, and the resulting information forms the a-priori information which will be fed to the next stage. For the second decoding stage, the soft metrics of the even symbols together with the a-priori information produced by the first decoding stage is interleaved and fed to the decoder2. At the output of this decoder, the a-priori information is calculated as in the first decoding stage before being de-interleaved and forwarded to the decoder1. Then the iteration process is repeated until a prescribed maximum number of iterations is researched. Finally, the a-posteriori information at the output of the decoder 2 is sent to a decision unit, resulting in an estimation of the transmitted systematic information bits [36]. The marked code is inverted so that the code, the unmarked codes output without any operation, can be restored as it is,

According to the principle of MDL (Minimum description length), the main task of the decoder is to calculate the transmission probability of different symbols under the received sampling condition. For u_k (k is time index) bit, the posterior probability is expressed as [37]:

$$P_r\{u_k = i/r\}, i = 0, 1 \tag{2.33}$$

The soft decision output of the information symbol is:

$$A(u_k) = \ln\left\{\frac{Pr\{u_k = 1/R\}}{Pr\{u_k = 0/R\}}\right\} \tag{2.34}$$

To facilitate the calculation of soft decision output, three probabilities are defined:

$$\begin{gathered}
\alpha_k^i(m), \beta_k(m), \gamma_k^i(R_k, m', m) i = 0, 1 \\
\alpha_k^i(m) = \frac{\Pr\{u_k = i, s_k = m, R_1^k\}}{\Pr\{R_1^k\}} = \Pr\left\{u_k = i, s_k = \frac{m}{R_1^k}\right\} \\
\beta_k(m) = \frac{\Pr\left\{\frac{R_{k+1}^N}{s_k} = m\right\}}{\Pr\left\{\frac{R_{k+1}^N}{R_1^k}\right\}} \\
\gamma_K^i(R_K, m', m) = \Pr\left\{u_K = i, R_K, s_K = \frac{m}{s_{K-1}} = m'\right\}
\end{gathered} \tag{2.35}$$

For M-ary symbol, the information bits and error correcting bits have been synthesized into a M-ary symbol to be modulated and transmitted. At the receiver after demodulation, only the soft decision information of the symbol can be provided, and the information bits cannot be completely independent. Therefore, in the MAP decoding, the system information and the error correction information are considered together, that is to say, they cannot be separated from each other in the decoding. Then, the final soft decision output can only be divided into two parts:

$$\Lambda_k(u_k) = L_{2k} + \ln \frac{\sum\limits_m \sum\limits_{m'} \gamma_k^i(y_{1k}, m', m) \alpha_{k-1}(m') \beta_k(m)}{\sum\limits_m \sum\limits_{m'} \gamma_k^0(y_{1k}, m', m) \alpha_{k-1}(m') \beta_k(m)} \tag{2.36}$$

In (2.36), the first item is the internal information, which is directly related to the external information transmitted by another decoding unit, and the second item is considered as external information, which will be transmitted to another decoding unit.

2.4.1.2 GFDM Principle

Figure 2.19 illustrates the principle of the transmitter and receiver of GFDM. Binary data is modulated into QAM format and divided into KM complex symbols. Each symbol dk[m] is spread across k sub-carriers and m time slots for transmission, where $k = 0\ldots K - 1$, $m = 0\ldots M - 1$ are given. The complex data symbols $d_k[m]$ are up-sampled by a factor N, resulting in:

$$d_k^N[n] = \sum_{m=0}^{M-1} d_k[m]\delta[n - mN], n = 0, \ldots, NM - 1 \tag{2.37}$$

Where $\delta[\cdot]$ is the Dirac function, and N is the up-sampling factor. Therefore, after up-sampled by N, digital pulse shaping and up-conversion, the transmitted GFDM signal can be expressed as:

$$x[n] = \sum_{m=0}^{M-1} \sum_{k=0}^{K-1} d_k^N[n] g_T[n], g_T[n] = g[(n - mN)|MN] e^{i2\pi\frac{kn}{N}} \tag{2.38}$$

Each sub-carrier signal $g_T[n]$ is a shifted version of a circular filter g[n], by mN and k/N in the time and frequency domain respectively.

At the receiver, GFDM demodulation performed in frequency domain, which is opposite to the modulation module, shown in Fig. 2.19. Suppose y[n] are the received signal, thus the resulting signal at $n = mN$ is given by the expression [38].

$$\tilde{d}_k[m] = \left(y[n]e^{-j2\pi\frac{kn}{N}}\right) \otimes g_{Rx}[n]\Big|_{n=mN} \tag{2.39}$$

Where $\otimes g_{Rx}[n]$ denotes circular convolution with matched filter relative to n which is essential for tail biting at the receiver. In frequency domain, the formula (2.39) will be calculated by FFT and IFFT conversion flexibly. After digital filtering, the signal is down-sampled, thus turning sample index into QAM symbol [39].

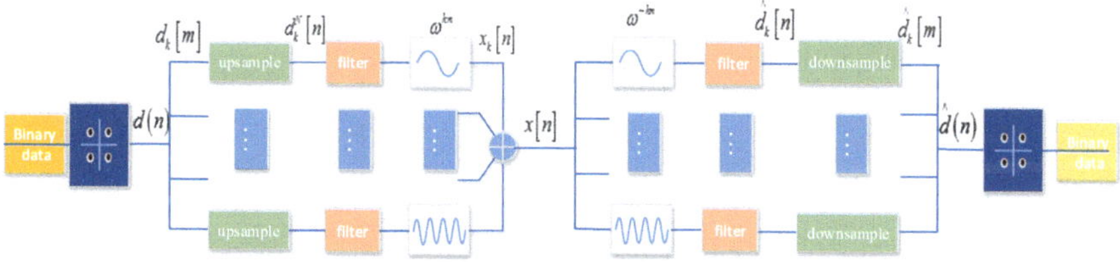

Fig. 2.19 The schematic diagram of GFDM signal

2.4.2 Conclusion

In this subsection, we have proposed a novel concatenated coded modulation based on GFDM for optical access system. The performance of the scheme combining the adaptive turbo code with trellis coded modulation (ATTCM) was successfully demonstrated in the experiment. The experimental results showing the concatenated coded modulation (ATTCM32QAM) can solve the error floor problem by providing 7.3 dB (BER = 1e-3) coding gain than uncoded 16QAM signal and achieves 2.3 dB (BER = 1e-3) performance improvement than that of trellis coded modulation 32QAM (TCM32QAM) after 1 km transmission. The results indicates that the designed novel concatenated coded modulation achieved high spectrum efficiency and significant coding gain, and the complexity is also reduced by integrating turbo code with trellis code modulation together. This work shows that the proposed method will be a promising solution for future 5G access system.

2.5 Probabilistic Shaped Trellis Coded Modulation

With the requirement for large capacity communication systems, the technology to close the gap between the achievable information rate and the Shannon capacity limit attracts more and more attention. In this section, we present a novel scheme of trellis coded modulation combined with probabilistic shaping (PS-TCM) in the intensity-modulation/direct detection (IM/DD) system, using generalized frequency division multiplexing (GFDM). The principle of PS-TCM is analyzed with mutual information. We experimentally demonstrate the proposed scheme in the optical GFDM system. The results show that 4 Gb/s PS-TCM-32QAM signal achieves ~2 dB gain over the regular 16QAM signal in the back-to-back case. After 20 km transmission, the scheme of PS-TCM-32QAM provides 1.8 dB performance gain over that of the regular 16QAM signal.

2.5.1 Principle of PS-TCM

The schematic configuration of channel model is shown in Fig. 2.20, which includes probabilistic shaped-trellis coded modulation (PS-TCM) encoder and GFDM

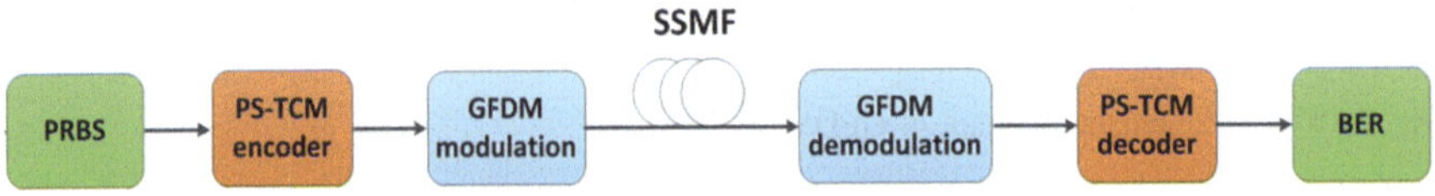

Fig. 2.20 Schematic diagram of system model

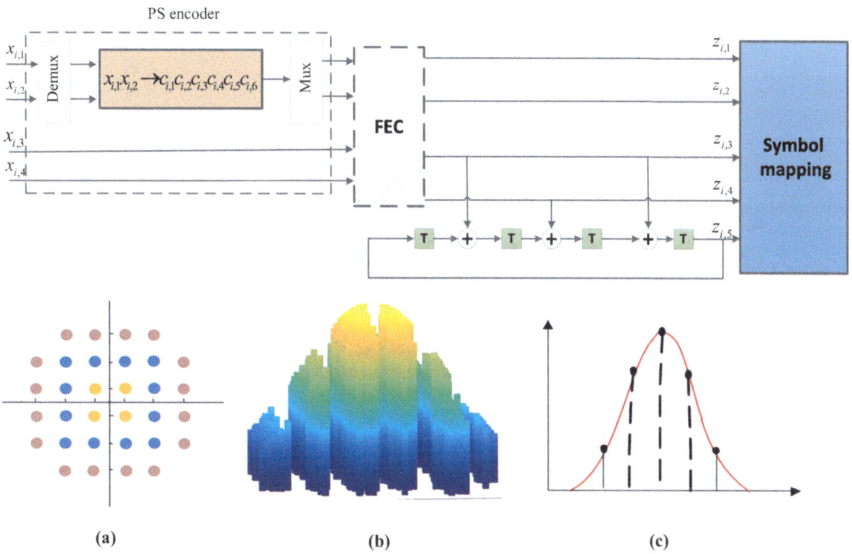

Fig. 2.21 The principle for PS-TCM32QAM

Table 2.1 Construction of PS encoder

$x_{i,1}x_{i,2}$	$x_{i,1}x_{i,2} \rightarrow c_{i,1}c_{i,2}c_{i,3}c_{i,4}c_{i,5}c_{i,6}$
00	000000
01	010000
11	000001; 000010; 000011; 000100; 000101; 000110; 000111; 001000; 001001; 001010; 001011; 001100; 001101; 001110; 001111; 010001; 010010; 010011; 010100; 010101; 010110; 010111; 011000; 011001; 011010; 011011; 011100; 011101; 011110; 011111; 100001; 100010; 100011; 100100; 100101; 100110; 100111; 101000; 101001; 101010; 101011; 101100; 101101; 101110; 101111; 110001; 110010; 110011; 110100; 110101; 110110; 110111; 111000; 111001; 111010; 111011; 111100; 111101; 111110; 111111
10	100000

modulation at the transmitter, as well as GFDM demodulation and PS-TCM decoder at the receiver. Each part is described in detail in the following subsections.

The principle of PS-TCM is shown in Fig. 2.21. The original bits are divided into four branches by serial-parallel conversion. Then the four branch signals are divided into two groups. One group is variable and denoted by $x_{i,1}, x_{i,2}$, which is responsible for PS with the mapping of different length of bits. The other group is constant with a fixed length and denoted by $x_{i,3}, x_{i,4}$, which is used for convolution coding of TCM. Here, one bit of redundancy check is added to design the TCM coding with a code rate of R = 4/5 [40]. PS encoder is realized by adding redundancy bits according to the symbol prefix [41]. The unequal constellation distribution is achieved with higher proportion of low-energy symbols and lower proportion of high-energy symbols, as shown in Fig. 2.21 a and b. The mapping rule is shown in Table 2.1,

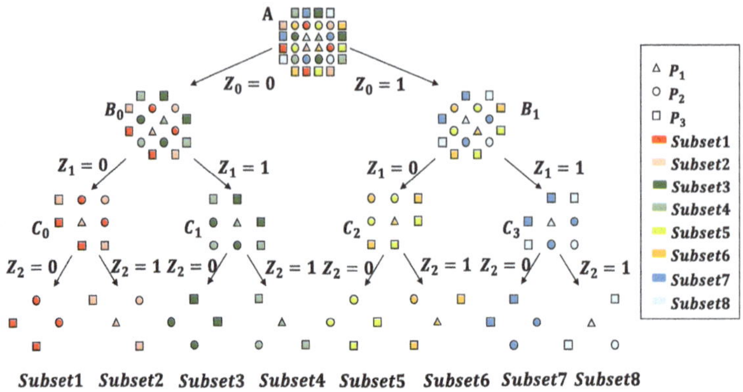

Fig. 2.22 Set partitioning process for PS-TCM32QAM

which adds bits with the source code as prefix. It can be seen that the probability of 0 and 1 before shaping is 50% and 50%, the probability of 0 after mapping in Table 2.1 is 66.7%. After PS encoder and TCM coding, the five bits are mapped into one symbol. Therefore, the expected probability distribution of the constellation points can be met by adjusting the mapping rule for the Gaussian channel, as shown in Fig. 2.21c. Moreover, the proposed scheme can also be integrated with FEC shown in Fig. 2.21 with the dotted boxes.

The main idea behind PS-TCM is set partitioning, which takes PS and TCM into consideration together to maximize the Euclidean distance between the points of the same subset, and it can improve the error correction ability by transmitting small bits with large constellation, without reducing the bandwidth and power utilization. PS technology is applied to optimize the probability distribution of symbols to increase the spectral efficiency (SE) and reduce the average transmitted energy. Both PS and TCM can be well compatible together. The set partitioning process for PS-TCM32QAM is shown in Fig. 2.22. Different shapes are used to present various probability distributions. That is, the probability of 4 constellation points in the innermost circle (represented by a triangle) is P1, the probability of the middle 12 constellation points (denoted by a circle) is P2, and the probability of 16 constellation points (represented by a square) in the outermost circle is P3. Here, $4 \times P1 + 12 \times P2 + 16 \times P3 = 1$. The constellation is divided into subsets in half and half, and each subset has an increasing distance step by step. After 3 steps the constellation will be finally divided into 8 subsets which are differentiated with different colors. From Fig. 2.22, it can be seen that considering PS with the constellation, the point in each subset is related to the probability distribution. When calculating the metrics with minimum Euclidean distance, the function will be multiplied by the probability coefficient and modified to be $L = \sum_{k-1}^{N_c} |\gamma_k - \rho b_k|^2$, where N_c is length of the coded sequence, γ_k is the received signal, b_k is the most likely transmitted sequence, and ρ is the probability coefficient. Therefore, probabilistic shaping and trellis coding are compatible together to improve the performance.

At the receiver side, Viterbi algorithm is used for TCM decoding, which brings complexity as well as gain. The complexity of the algorithm is proportional to the length of transmitted symbols and the number of trellis states [42, 43]. In practice, in order to make Viterbi decoding be more suitable for optical network, some researchers have focused on the methods to reduce the complexity of Viterbi algorithm by reducing the number of states [44, 45]. However, the complexity reduction is achieved using the M-algorithm, which will introduce the performance damage [45, 46]. Therefore, considering both the sensitivity gain and complexity of convolutional encoder, 16-state encoder architecture is selected in the system. In practical application, the complexity can be reduced by reducing the number of states under the condition of the practical demand.

GMI is regarded as an effective approach to represent the capacity of the communication system [47, 48]. As a function of SNR, the achieved GMI for different modulation formats including 16QAM, TCM-32QAM, PS-16QAM and PS-TCM-32QAM is shown in Fig. 2.4. It can be seen that when SNR $<$ 13 dB, the GMI of 16QAM is better than that of TCM-32QAM because the code gain of TCM is small in low SNR region. While the SNR increases to be larger than 13 dB, GMI of TCM-32QAM will converge towards 5bits/symbol, and that of 16QAM will towards 4bit/ symbol, that's because TCM-32QAM contains one bit of redundancy check. Similarly, When SNR$>$12.5 dB, the GMI of PS-TCM-32QAM with entropy of 4.62 is better than that of PS-16QAM because trellis code brings coding gain. It is observed that the GMI is improved significantly with shaping and coding gain below a certain SNR after employing PS and TCM.

The data rate of the proposed scheme can be calculated by R = H (ρ)-m (1-r) [47, 49], where H (ρ) is the entropy of the PS-TCM-32QAM, m is the number of bits per symbol and r is the coding rate. In our proposed scheme, the code rate r of TCM is 4/5.

Based on the previous theoretical analysis, the formula of GMI is expressed by [47, 48].

$$GMI \approx -\sum_{x\in\chi} P_X(x)\log_2 P_X(x) + \frac{1}{N}\sum_{k=1}^{N}\sum_{i=1}^{m}\log_2\frac{\sum_{x\in\chi_{k,j}} q_{\frac{Y}{X}}(y_k x)P_X(x)}{\sum_{x\in X} q_{\frac{Y}{X}}(y_k x)P_X(x)} \tag{2.40}$$

Where $P_X(x) = \frac{1}{\sum_{k=1}^{M} e^{-w_k^2}} e^{-w^2}$ is the form of Maxwell-Boltzmann probability mass function with scaling factor ≥0. $q_{Y/X}(y/x)$. is the function of channel conditional transfer probability and it can be presented by

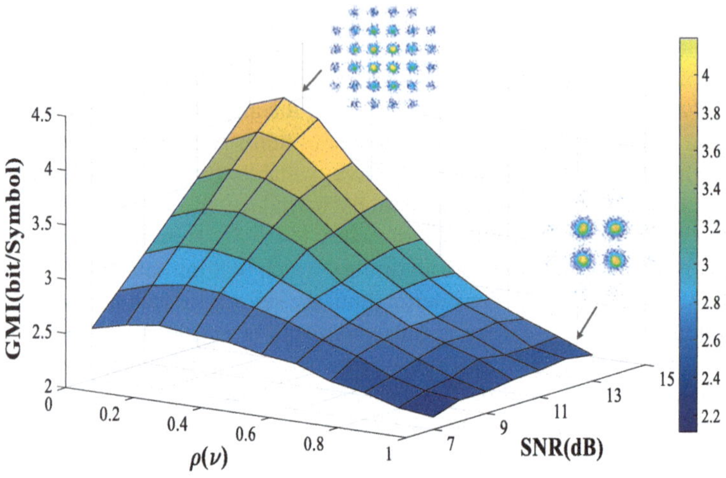

Fig. 2.23 3-D diagram for PS-TCM-32QAM

$$q_{\frac{Y}{X}}\left(\frac{y}{x}\right) = \frac{1}{\sqrt{2\pi\sigma^2}} e^{-\frac{(y-x)^2}{2\sigma^2}} = \frac{1}{\sqrt{2\pi\sigma^2}} e^{-\frac{n^2}{2\sigma^2}} = p(n) \tag{2.41}$$

From Eqs. (2.40) and (2.41), we can observe that the GMI is related to the parameter ν and σ, in which ν decides the probability mass function and σ is the noise variance of channel.

For PS-TCM32QAM signal, the required SNR can be written as [50]

$$SNR \triangleq \frac{E\left(\rho |X|^2\right)}{\gamma \cdot \sigma^2} \tag{2.42}$$

Where ρ is the coefficient related to probability distribution, γ represents the gain factor introduced by trellis coding. From Eqs. (2.40) and (2.42), we find that the GMI is closely related to SNR and the parameter ν. Figure 2.23 presents the 3-D diagram of the GMI for PS-TCM-32QAM, which shows that the GMI is directly proportional to the SNR, and inversely proportional to ν.

2.5.2 *Principle of GFDM Modulation*

The principle of GFDM modulation is shown in Fig. 2.24, which is a multi-carrier modulation scheme using non-rectangular pulse shaping filter. The input signal is unsampled by a factor of N, and the lowpass pulse shaping filter (LPF), tail biting are applied to shape the signal to reduce band radiation. After subcarrier up-conversion and digital-to-analog conversion, the output samples can be expressed as [24–26]

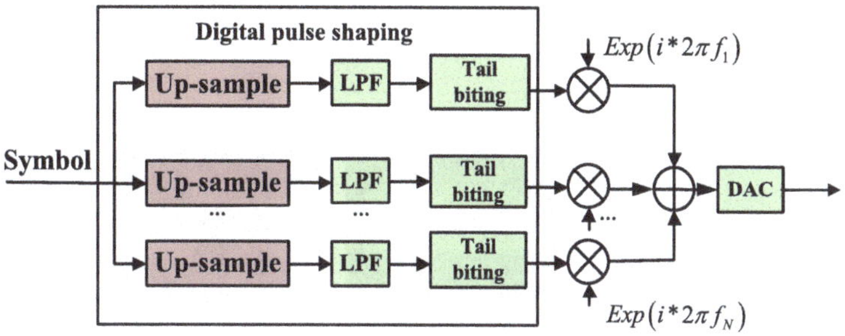

Fig. 2.24 The principle of GFDM

$$y_n(k) = [x(n,k) * g(k)] \bullet e^{j2\pi k f_n} \tag{2.43}$$

Where $x(n,k)$ is the transmitted symbol on the nth carrier and k^{th} time slot, g(k) is the pulse shaping filter and f_n indicates different carriers' frequency.

At the receiver, the GFDM signal is converted to digital baseband signal, and then demodulated by matched filter (MF) receiver with frequency domain equalization [25].

The scheme of PS-TCM based on GFDM provides significant performance gain, thus increasing transmission distance and approaching the high data rate demand that the optical fiber system requires, which enables the system to support a huge number of optical network units. There is no research reported about PS-TCM based on GFDM for data center optical network until now. It is a promising candidate to optimize the PS-TCM in GFDM system, which brings a better trade-off between effectiveness and reliability performance for the optical fiber communication system.

2.5.3 Experimental Setup and Results

Figure 2.25 shows the experimental setup for PS-TCM-32QAM optical system. At the transmitter, the lightwave is generated from an external cavity laser (ECL) with the line-width of 100 kHz. The central wavelength of the ECL is 1549.480 nm and its output power can be adjusted from 7.5 dBm to 15 dBm. The Mach-Zehnder modulator (MZM) with 40 GHz bandwidth is adopted to modulate the electrical signal onto the lightwave. The Arbitrary Waveform Generator (AWG) at 25GSa/s with 3 dB bandwidth of 14 GHz is used to produce the electrical signal (PS-TCM-32QAM GFDM) with a bit rate of 4 Gbit/s, which is then sent into the radio-frequency (RF) port of the MZM biased at 5.4V. Then the optical signal is amplified by a commercial Erbium-doped fiber amplifier (EDFA) before it is launched into 20 km standard single mode fiber (SSMF) with a total loss of 4.5 dB. The variable optical attenuator (VOA) is applied to change the received power of the photodetector (PD). The optical filter is used to suppress the noise out

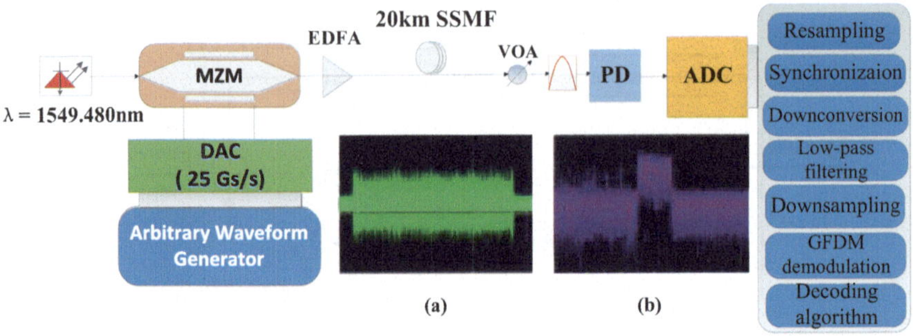

Fig. 2.25 Experimental setup. (**a**) The detected electrical signal in time domain. (**b**) The electrical spectrum of the detected signal. *DAC* digital-to-analog converter, *MZM* Mach-Zehnder modulator, *EDFA* Erbium-doped fiber amplifier, *VOA* variable optical attenuator, *PD* photodetector, *ADC* analog-to-digital converter

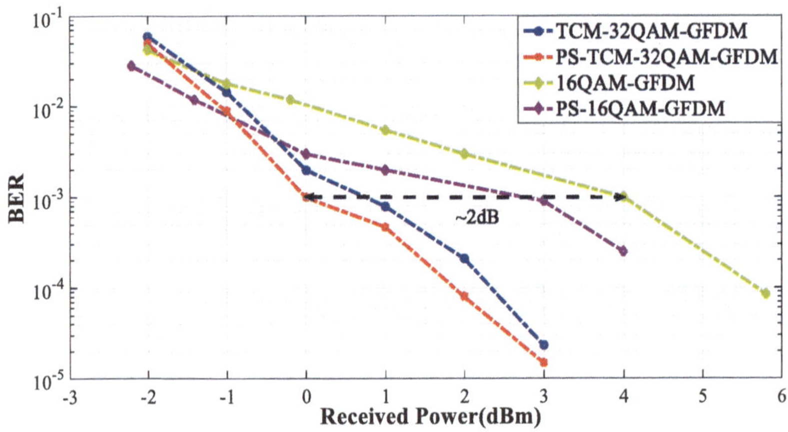

Fig. 2.26 BER performance for mQAM-GFDM signal

of band. Finally, the signal is sampled by an analog to digital converter (ADC) with 50 GSa/s for the subsequent offline DSP. The received signal corresponding to the time domain and frequency domain is shown in Figs. 2.25 a and b, respectively. The offline DSP includes resampling, synchronization, down-conversion, low-pass filtering, down-sampling, MF algorithm, and decoding algorithm. The MF algorithm is used for GFDM demodulation, and the Viterbi algorithm is used for TCM decoding.

In the experiment, we have measured the bit error ratio (BER) performance for the 16QAM, TCM-32QAM, and PS-TCM-32QAM signals in GFDM system at the same bit rate of 4 Gb/s. Figure 2.26 illustrates the BER curve versus different received power. It can be seen that the PS combined with TCM brings the performance gain of about ~2 dB compared to 16QAM at BER of 1×10^{-3}, whereas the PS brings a gain of about 0.7 dB for TCM-32QAM signal.

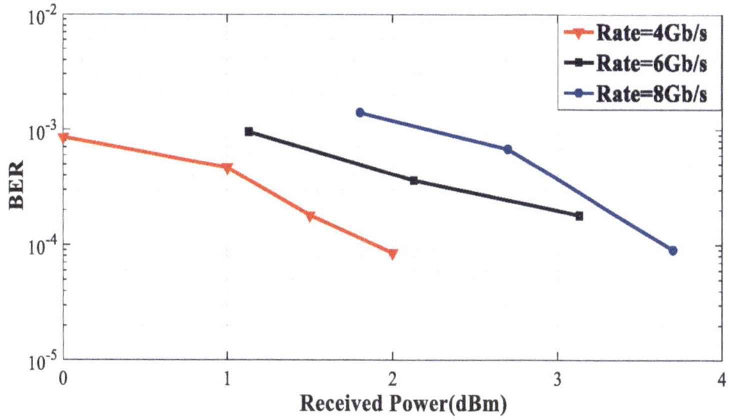

Fig. 2.27 BER versus received power at different rates for PS-TCM-32QAM

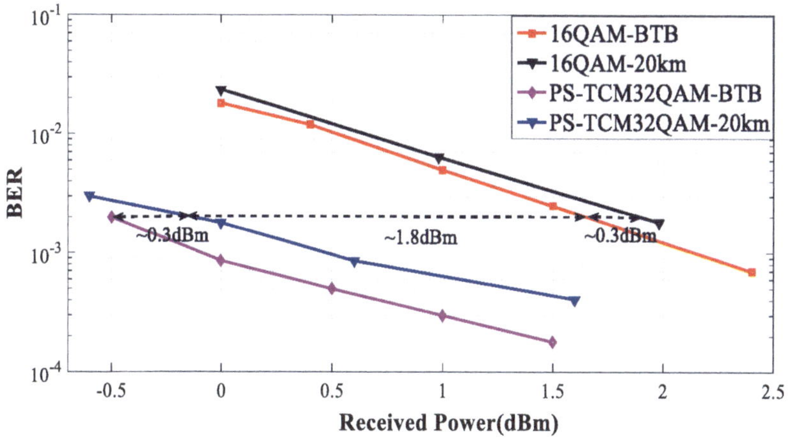

Fig. 2.28 BER versus received power at different distance for 16QAM and PS-TCM-32QAM

Figure 2.27 presents the BER performance for PS-TCM-32QAM with bit rates of 4Gb/s, 6Gb/s, and 8Gb/s, respectively. When the rate is increased by 2Gb/s, ~1 dB power penalty will be introduced for the system at the BER of 1×10^{-3}. It can be considered that when the data rate is higher, the more noise components in the received signal and the BER performance of the signal is worse with the same received optical power.

Figure 2.28 shows the BER performance versus the received power at different distance for the 16QAM and PS-TCM-32QAM signals, respectively. The bit rate is fixed at 4 Gb/s. It can be seen that 20 km fiber transmission causes ~0.3 dB power penalty (which is caused from the nonlinearity of the transmission link or devices) for both the PS-TCM-32QAM and 16QAM signal. The PS-TCM-32QAM signal provides a performance improvement of 1.8 dB compared to the 16QAM signal after 20 km transmission.

2.6 Summary

This chapter presents a comprehensive suite of state-of-the-art modulation and coding strategies specifically designed for next-generation elastic optical networks. It commences with an adaptive non-uniform PSK modulation scheme that dynamically optimizes constellation amplitudes to balance receiver simplicity with service-specific quality-of-service (QoS) requirements. Building upon adaptive principles, a three-dimensional hierarchical QAM framework is introduced, employing multi-dimensional constellation reconfiguration to improve error resilience and receiver sensitivity. To further advance spectral efficiency, a probabilistically shaped (PS) 4D modulation scheme is demonstrated within a self-homodyne coherent transmission architecture. Furthermore, a hybrid concatenated coding scheme combining adaptive turbo codes with trellis-coded modulation (ATTCM) is implemented in a generalized frequency division multiplexing (GFDM) framework, achieving robust coding gains and spectral efficiency while maintaining low implementation complexity. Finally, we propose a novel probabilistic shaping trellis-coded modulation (PS-TCM) scheme based on GFDM for data center optical interconnect systems. The achievable benefits of PS-TCM-32QAM are discussed by comparing it with the regular 16QAM and TCM-32QAM both theoretically and experimentally.

References

1. R. Van Der Linden, N.C. Tran, E. Tangdiongga, et al., Increasing flexibility and capacity in real PON deployments by using 2/4/8-PAM formats. IEEE J. Opt. Commun. Netw. **9**, A1–A8 (2017)
2. M. Yoshida, T. Hirooka, K. Kasai, et al., Real-time adaptive 4–64QAM, 20–60 Gbit/s quantum noise stream cipher transmission over 320 km with FPGA-based transmitter and receiver, in *2015 European Conference on Optical Communication*, (IEEE, 2015)
3. M. Isaka, H. Imai, Hierarchical coding based on adaptive multilevel bit-interleaved channels, in *IEEE Vehicular Technology Conference, 2000*, (IEEE, 2000)
4. M.B. Pursley, J.M. Shea, Adaptive nonuniform phase-shift-key modulation for multimedia traffic in wireless networks. IEEE J. Sel. Areas Commun. **18**, 1394–1407 (2000)
5. Z. Yue, Q. Zhang, X. Xin, W. Zhang, X. Pan, Q. Tian, Y. Tao, F. Tian, X. Liu, Trellis coded modulation with three-dimension 9QAM mapping based on a novel probabilistic shaping method. Opt. Commun. **449**, 45–52 (2019)
6. X. Chen, Z. Feng, M. Tang, B. Li, H. Zhou, S. Fu, D. Liu, Three-dimensional adaptive modulation and coding for DDO-OFDM transmission system. IEEE Photon. J. **9**(2), 1–21 (2017)
7. M. Che, T. Kuboki, K. Kato, Trellis coded three dimensional carrierless amplitude and phase modulation, in *OECC/PSC 2019, TuP4-B10*, (IEEE, 2019)
8. J. Ren, B. Liu, X. Wu, L. Zhang, Y. Mao, X. Xu, Y. Zhang, L. Jiang, J. Zhang, X. Xin, Three-dimensional probabilistically shaped CAP modulation based on constellation design using regular tetrahedron cells. J. Lightwave Technol. **38**(7), 1728–1734 (2020)
9. H. Mukhtar, M.E. Tarhuni, An adaptive hierarchical QAM scheme for enhanced bandwidth and power utilization. IEEE Trans. Commun. **60**, 2275–2284 (2012)

10. T. Yoshida, M. Karlsson, E. Agrell, Hierarchical distribution matching for probabilistically shaped coded modulation. J. Lightwave Technol. **37**(6), 1579–1589 (2019)
11. B. Liu, L.J. Zhang, X.J. Xin, et al., Flexible generation of optical nonuniform bit-mapping signal based on InP transmitter module. Opt. Lett. **39**, 220–223 (2014)
12. M.O. Zamkotsian, K.P. Peppas, F. Lazarakis, P.G. Cottis, Multilevel spatial hierarchical modulation: An efficient scheme for unequal error protection underrician fading. IEEE Trans. Veh. Technol. **64**(11), 5177–5186 (2015)
13. H. Kim, Y. Shang, S. Kim, T. Jung, New hierarchical modulation scheme using a constellation rotation method, in *2016 IEEE International Conference on Signal and Image Processing, ICSIP, Beijing*, (IEEE, 2016), pp. 703–707
14. M. Dalla Santa, C. Antony, G. Talli, P.D. Townsend, Power budget improvement in passive optical networks using PAM4 hierarchical modulation. IEEE Photon. Technol. Lett. **29**(20), 1747–1750 (2017)
15. H. Sun, Y. Shen, S.X. Ng, L. Hanzo, Turbo Trellis coded hierarchical modulation for cooperative communications, in *2013 IEEE Wireless Communications and Networking Conference (WCNC), Shanghai*, (IEEE, 2013), pp. 2789–2794
16. H. Sun, S.X. Ng, L. Hanzo, Turbo trellis-coded hierarchical-modulation assisted decode-and-forward cooperation. IEEE Trans. Veh. Technol. **64**(9), 3971–3981 (2015)
17. H. Sun, S.X. Ng, C. Dong, L. Hanzo, Decode-and-forward cooperation-aided triplelayer turbo-trellis-coded hierarchical modulation. IEEE Trans. Commun. **63**(4), 1136–1148 (2015)
18. F. Tian et al., Design of probabilistic shaping 4D Ultra high order modulation format with 8APSK pilot aided carrier phase recovery. J. Lightwave Technol. **40**(12), 3688–3697 (2022)
19. A. Alex, F. Tobias, C. Bin, M.J.W. Frans, et al., Achievable information rates for fiber optics: applications and computations. Comput. Res. Reposit. **6**(2), 424–439 (2017)
20. X. Guo, Q. Wang, X. Li, L. Zhou, L. Fang, First demonstration of OFDM ECDMA for low-cost optical access networks. Opt. Lett. **40**, 2353–2356 (2015)
21. Y. Fang, J. Yu, J. Zhang, N. Chi, J. Xiao, G.-K. Chang, Ultrahigh-capacity access network architecture for mobile data backhaul using integrated W-band wireless and free-space optical links with OAM multiplexing. Opt. Lett. **39**, 4168–4171 (2014)
22. N. Cvijetic, OFDM for next-generation optical access networks. J. Lightwave Technol. **30**(4), 384–398 (2012)
23. B. Liu, L. Zhang, X. Xin, J. Yu, Robust generalized filter bank multicarrier based optical access system with electrical polar coding. IEEE Photon. J. **8**(5) (2016) Art. no. 7906507
24. L. Zhang, B. Liu, X. Xin, Secure optical GFBMC system based on cubic constellation masked method. Opt. Lett. **40**, 2711–2714 (2015)
25. A. Ortega, L. Fabbri, V. Tralli, Performance evaluation of GFDM over nonlinear channel, in *Proceedings of the International Conference on Information Communication Technology and System*, (2016), pp. 12–17
26. C. Browning, A. Farhang, A. Saljoghei, N. Marchetti, 5G wireless and wired convergence in a passive optical network using UF-OFDM and GFDM, in *Proceedings of the IEEE International Conference on Communications Workshops*, (2017), pp. 386–392
27. F. Chang, K. Onohara, T. Mizuochi, Forward error correction for 100 G transport networks. IEEE Commun. Mag. **48**(3), S48–S55 (2010)
28. I.B. Djordjevic, H.G. Batshon, LDPC-coded OFDM for heterogeneous access optical networks. IEEE Photon. J. **2**(4), 611–619 (2010)
29. L. Zhang, B. Liu, X. Xin, Y. Wang, Joint robustness security in optical OFDM access system with Turbo-coded subcarrier rotation. Opt. Exp. **23**(1), 13–18 (2015)
30. D. Liu, *Principle and application technology of turbo code [M]* (Publishing House Electron. Ind, Beijing, China, 2004)
31. M. Magarini, R. Essiambre, B. Basch, A. Ashikhmin, G. Kramer, Concatenated coded modulation for optical communications systems. Photon. Technol. Lett. **22**(16), 1244–1246 (2010)
32. H. Bülow, G. Thielecke, F. Buchali, Optical trellis-coded modulation (OTCM), in *Proceedings of Optical Fiber Communication Conference, Los Angeles, CA, USA*, (2004) paper WM5

33. H. Zhao, E. Agrell, M. Karlsson, Trellis-coded modulation in PSK and DPSK communications, in *Proceedings of the European Conference on Optical Communications, Göteborg, Sweden*, (2006), pp. 1–2
34. M.S. Kumar, H. Yoon, N. Park, Performance evaluation of trellis code modulated oDQPSK using the KLSE method. IEEE Photon. Technol. Lett. **19**(16), 1245–1247 (2007)
35. Q. Yang, Y. Ma, W. Shieh, 1 Tbit/s single-channel coherent optical OFDM transmission with trellis-coded modulation. Electron. Lett. **45**(20), 1045–1047 (2009)
36. S. Alreesh, C. Schmidt-Langhorst, R. Emmerich, P. Berenguer, C. Schubert, J.L. Fisher, Four-dimensional trellis coded modulation for flexible optical communications. J. Lightwave Technol. **35**(2), 152–158 (2017)
37. J. Wang, X. Wang, Y. Dai, Z. Zheng, Study of modulation techniques for combing turbo code with TCM. Comput. Netw. **8**, 24–25 (2000)
38. N. Michailow, S. Krone, M. Lentmaier, G. Fettweis, Bit error rate performance of generalized frequency division multiplexing, in *Proceedings of the IEEE Vehicular Technology Conference*, (2012), pp. 1–5
39. N. Michailow, I. Gaspar, S. Krone, M. Lentmaier, Generalized frequency division multiplexing analysis of an alternative multi-carrier technique for next generation cellular systems, in *Proceedings of the International Symposium on Wireless Communication Systems*, (2012), pp. 171–175
40. N. Palgy, R. Zamir, Dithered probabilistic shaping, in *2012 IEEE 27th Convention of Electrical and Electronics Engineers in Israel, Eilat, Israel*, (2012), pp. 1–5
41. G. Bocherer, P. Schulte, F. Steiner, Probabilistic shaping and forward error correction for fiber-optic communication systems. J. Lightwave Technol. **37**(2), 230–243 (2019)
42. C. Prodaniuc, N. Stojanovic, F. Karinou, Z. Qiang, R. Llorente, Performance comparison between 4D trellis coded modulation and PAM-4 for low-cost 400 Gbps WDM optical networks. J. Lightwave Technol. **34**(22), 5308–5316 (2016)
43. H.C. Myburgh, L.P. Linde, Reduced complexity combined soft-decision MLSE equalization and decoding, in *Australasian Telecommunication Networks & Applications Conference*, (IEEE, 2008)
44. Y. Fu, M. Bi, D. Feng, X. Miao, H. He, W. Hu, Spectral efficiency improved 2D-PAM8-Trellis coded modulation for short reach optical system. IEEE Photon. J. **9**(4) (2017)
45. Y. Fu, D. Feng, M. Bi, H. Xin, K. Zhang, M. Zhang, H. He, W. Hu, 25Gbps Two-dimensional trellis coded PAM4 TDM-PON transmission based on 10G optics, in *Opto-electronics & Communications Conference*, (IEEE, 2017)
46. M. Zamani, H. Najafi, D. Yao, J. Mitra, C. Li, Z. Zhang, Trellis-based feed-forward carrier recovery for coherent optical systems. Opt. Express **24**(20), 23531–23542 (2016)
47. S.L. Zhang, F. Yaman, Design and comparison of advanced modulation formats based on generalized mutual information. J. Lightwave Technol. **36**(2), 416–423 (2017)
48. I.F.D.J. Ruiz, A. Ghazisaeidi, O.A. Sab, P. Plantady, A. Calsat, S. Dubost, L. Schmalen, V.T. Letellier, J. Renaudier, 25.4 Tb/s Transmission over trans-pacific distances using truncated probabilistically shaped PDM-64QAM. J. Lightwave Technol. **36**(6), 1354–1361 (2018)
49. T. Fehenberger, R. Maher, A. Alvarado, P. Bayvel, N.T. Hanik, Sensitivity gains by mismatched probabilistic shaping for optical communication systems. Photon. Technol. Lett. **28**(7), 786–789 (2016)
50. T. Fehenberger, A. Alvarado, G. Bocherer, N. Hanik, On the impact of probabilistic shaping on SNR and information rates in multi-span WDM system, in *Proceedings of the Optical Fiber Communication Conference, Los Angeles, CA, USA*, (Optica Publishing Group, 2017), pp. 1–3. paper.M3C.4

Chapter 3
Multi-carrier Generation and Transmission Technology

With the increasing demand of data services in cloud computing, high-definition video, and big data applications, the Tb/s optical transmission with multicarrier technologies has become a worldwide research hotspot for high-speed long haul optical network, and broad-band optical access network as well [1–5]. There are some theoretical and experimental studies about generating multicarrier by using the typical cascaded Mach–Zehnder modulators (MZMs) and phase modulators [6–8]. However, these schemes need a high-power radio frequency amplifier to obtain a large number of multi-carriers. Similarly multicarrier based single sideband (SSB) modulator using recirculating frequency shifter (RFS) has been demonstrated extensively [9–14]. Between these two methods, the latter has the advantages of lower drive voltage and the ability to generate larger number of carriers. The stability of SSB modulator based RFS loop due to the nonlinearity of optical I/Q modulator (IQM) has been studied in [11] under the perfect implementation. However, there exist other parameters such as the deviation from right bias voltage and unbalanced drive signal, which affect the entire output properties under the imperfect implementation in practice [11]. Recently, a double-side RFS structure has been proposed which is capable of generating multiple carriers [15, 16]. The scheme to generate a large number of optical coherent and frequency-locked multicarrier utilizes two RFS loops based on different frequency shifting direction SSB modulation in IQMs. The tone-to-noise ratio (TNR) performance with the effect of polarization state have been theoretically analyzed and experimentally investigated [16]. Jianping Li reported the scheme to generate hundreds of carriers based on a complementary frequency shifter with double RFS loops with only one I/Q modulator, as a result, the cost of the proposed configuration can be reduced further, and the principle can be described with more details. The generated 100 carriers are not flat (with a flatness of ~5 dB) at the two sides, whereas the SNR $<$ 40 dB [15]. In addition, there are other theoretical and experimental studies about generating multicarrier. In [17, 18] the authors deployed the cascaded phase modulators and also used RFS loop based on EDFA for the generation of coherent and frequency locked optical multi-carrier. The performance of 113 coherent and frequency-locked subcarriers with tone-to-noise

F. Tian, X. Xin, *High-Speed Large Capacity Optical Fiber Communications*,
Wireless Networks, https://doi.org/10.1007/978-3-031-94211-2_3

ratio larger than 26 dB and amplitude difference of 5 dB obtained after a tilt filter covering totally 22.6 nm has been reported in [19]. Using this novel scheme, the authors experimentally demonstrated 11.2-Tb/s optical OFDM PDM-QPSK signal transmission over 640-km SMF-28. Furthermore, based on a directly modulated laser (DML) and a phase modulator (PM) in cascade through synchronous sinusoidal radio frequency (RF) signal, the generated 16 optical subcarriers have frequency spacing of 12.5 GHz and power difference of <3 dB in [20]. Another novel method utilized three-branch-waveguide-interferometer-based single-side-band modulator [21]. The simulation results show that the TBWI-based MCG is able to generate multiple frequency-locked carriers.

In this subsection, we introduce a novel double sideband RFS structure based on DP-MZM. The principle used in the current study is theoretically analyzed in detail with simulation results. Relatively high flatness (0.7769 dB, 1.148 dB separately), SNR $>$ 40 dB than the traditional SSB based on RFS architecture [10] is obtained, and the efficiency of proposed multi-carrier generation technique is increased. And, we have successfully generated 50 stable frequency-locked carriers. The operating principle of multicarrier generation is analyzed and some important factors that affect the stability of the multicarrier are experimentally investigated. These impact factors include the imbalanced performance of the in phase/quadrature (I/Q) modulator, the accuracy of phase difference and the equality of the powers between two RF signals, the enough gain of the optical amplifier, the bandwidth and sharp window spectrum of the tunable filter, and the stability of the optical signals' states of polarization (SOP).

3.1 Theoretical Analysis

3.1.1 Analysis of the Structure and Theory of Recirculating Frequency Shifter (RFS)

Figure 3.1 shows the configuration of the RFS consisting of a closed loop, which comprises a 50:50 coupler, an I/Q modulator, an optical amplifier to compensate the frequency conversion loss and a tunable bandpass filter with flat-top shape to select the number of tones required, and a manual polarization controller to realize the adaptive SOP control and stabilization of signals.

The optical I/Q modulator consists of two Mach Zehnder modulators (MZM) in parallel configuration, shown in Fig. 3.2. There are two RF ports (RF A and RF B) to drive two MZMs, and three direct current (DC) biases in which two (DC A and DC B) are used for MZMs, and one (DC C) is used to generate the $\pi/2$ phase shift between the two MZMs. To observe the symmetry of the I/Q modulator, the power transfer curves of the two MZMs are measured as shown in Fig. 3.3, which shows some differences with the variation of DC A port and DC B port; RF A and RF B are fed with two equal powers but $\pi/2$ phase shifted clock signals. The precise clock

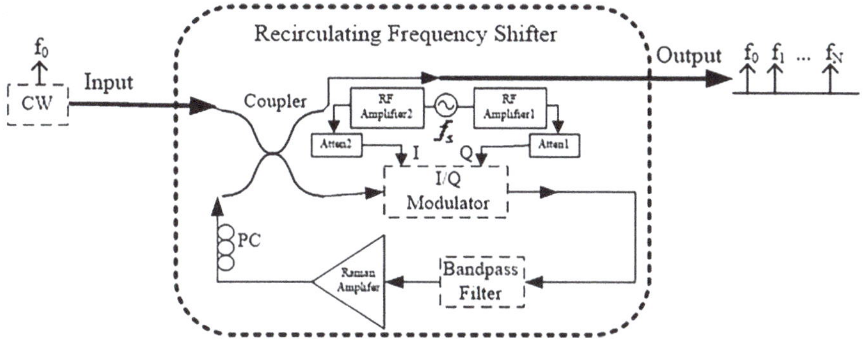

Fig. 3.1 Schematic of the RFS as a multicarrier generator

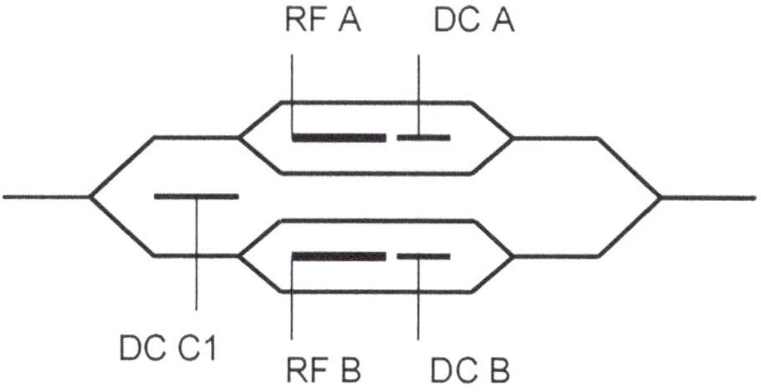

Fig. 3.2 Schematic diagram of the I/Q modulator

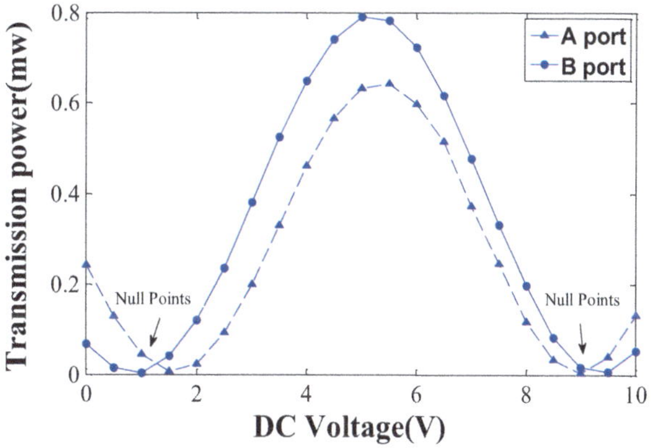

Fig. 3.3 Transmission curves of the I/Q modulator

signals are obtained from the signal generator, two high-power RF amplifiers and two adjustable attenuators are used to ensure the equal power of the RF inputs, and the $\pi/2$ phase difference can be obtained by adjusting the time delay of the outputs of signal generator.

The continuous-wave laser is used as the seed optical source of the SSB modulator-based RFS. The input optical field is represented by $E_{in} = E_0\, exp\,(j2\pi f_0 t)$, where f_0 represents the seed frequency. The cosine and sine signals with the frequency f_m (12.5 GHz) are injected into RF A port and RF B port of the I/Q modulator respectively. The output E-field of I/Q modulator after the first-round trip (RT) can be approximately expressed as follows:

$$\begin{aligned} E_{\text{out}} &\propto E_0 \exp(j2\pi f_0 t)[\cos(2\pi f_m t) + \text{j}\sin(2\pi f_m t)] \\ &= \text{K}\exp[j2\pi(f_0 + f_m)t] \end{aligned} \tag{3.1}$$

where K is the amplitude of the output. We can see that the first shifted frequency tone $f_1 = f_0 + f_m$has produced. Then f_1 tone is split into two branches; one coupled out and the other recirculating back to the input of the I/Q modulator. So, the f_2 tone with its frequency of f_0 + $2f_m$ will be generated after the second RT. Likewise, after the Nth round, we will have f_N tone shifted from the previous f_{N-1} tone. With continued recirculating in the loop, a series of frequency tones $f_2 = f_0 + 2f_m$, $f_3 = f_0 + 3f_m$, …, $f_N = f_0 + Nf_m$will be produced. However, the generated tones with the frequencies higher than f_N will be filtered out by adjusting the center wavelength and bandwidth of the optical bandpass filter placed in the loop.

Considering the input signal as $E_{in}(t) = E_0\ exp\ (j2\pi f_0 t)$, and RF drive signals as $f_I(t) = V_{pp}\ sin\ (2\pi f_m t), f_Q(t) = V_{pp}\ cos\ (2\pi f_m t)$. Assuming the total number of desired tones number is N + 1, the output of the SSB-based RFS can be denoted $f_0, f_1, \cdots, f_N$. The output electronic field of I/Q modulator can be expressed as follows under the ideal operation conditions:

$$E_{\text{out}}(t) = \frac{E_{\text{in}}(t)}{2}\left[\text{j}\sin\left(\frac{\pi}{2}\frac{f_I(t)}{V_\pi}\right) + \sin\left(\frac{\pi}{2}\frac{f_Q(t)}{V_\pi}\right)\right] \tag{3.2}$$

when using the Jacobi--Anger expansion, (3.2) can be expressed as

$$\begin{aligned} E_{\text{out}}(t) = \frac{E_{\text{in}}(t)}{2} &\cdot [\text{j}\sin(\delta_m \sin(2\pi f_m t))) \\ &+ \sin(\delta_m \cos(2\pi f_m t))]\exp(j\phi_{\text{RT}}) \\ &= E_{\text{in}}(t) \cdot j\sum_{k=1}^{\infty} J_{2k-1}(\delta_m) \\ &\times \sin[2\pi(2k-1)f_m t]\exp(j\phi_{\text{RT}}) \\ &+ E_{\text{in}}(t) \cdot \sum_{k=1}^{\infty} j^{2k-2} J_{2k-1}(\delta_m) \\ &\times \cos[2\pi(2k-1)f_m t]\exp(j\phi_{\text{RT}}) \\ &= E_{\text{in}}(t) \cdot [J_1(\delta_m)\exp(j2\pi f_m t) - J_3(\delta_m) \\ &\times \exp(-j6\pi f_m t)]\exp(j\phi_{\text{RT}}) \\ &+ E_{\text{in}}(t) \cdot [J_5(\delta_m) \\ &\times \exp(j10\pi f_m t) - \cdots]\exp(j\phi_{\text{RT}}) \end{aligned} \tag{3.3}$$

where $\delta_m = (\pi V_{pp})/(2V_\pi)$ denotes the phase modulation depth, $J_{2k-1}(\delta_m)$ are the first kind odd-order Bessel functions, ϕ_{RT} is the phase delay per RT. In most cases, the third-order harmonic is dominant among all the harmonics and ultimately affects the output property of the RFS. Ignoring all the high-order harmonics beyond the third order, the output of the RFS after the first RT can be expressed as

$$\begin{aligned} E_1(t) = E_{\mathrm{in}}(t) + E_{\mathrm{in}}(t) \cdot J_1(\delta_m)[\exp(j2\pi f_s t)) \\ +b\exp(-j6\pi f_s t)]\exp(j\phi_{\mathrm{RT}}) \end{aligned} \tag{3.4}$$

where $b = -J_3/J_1$ stands for the crosstalk coefficient which depends on the modulator drive voltage V_{pp}, and it affects the flatness and stability of the multicarrier. Besides, we have discussed some imbalance factors that influence the performance of the multicarrier. They include: (1) the balance imperfection of the I/Q modulator; and (2) the imbalance of RF clock signal. Considering the RF drive signals as $f_I(t) = V_{bI} + V_{ppI} \cos(2\pi f_m t)$, $f_Q(t) = V_{bQ} + V_{ppQ} \sin(2\pi f_m t + \Delta\theta)$, and $f_{PM}(t) = V_{bPM}$. The normalized transfer function of the I/Q modulator can be given by

$$\begin{aligned} T = \frac{1}{4}\{(1+\delta)[\exp(j\phi_{I1}) + \exp(-j\phi_{I2})] \\ +\exp(j\phi_{\mathrm{PM}})\left[\exp(j\phi_{Q1}) + \exp(-j\phi_{Q2})\right]\}\exp(j\phi_{\mathrm{RT}}) \end{aligned} \tag{3.5}$$

$$\begin{aligned} \phi_{I1} &= \frac{\pi}{2}\frac{\left[(V_{\pi I} + \Delta V_{bI}) + V_{\mathrm{pp}I}\cos(2\pi f_m t)\right]}{V_{\pi I}} \\ \phi_{I2} &= \frac{\pi}{2}\frac{\left[(V_{\pi I} + \Delta V_{bI}) + (1+\Delta V_I)V_{\mathrm{pp}I}\cos(2\pi f_m t)\right]}{V_{\pi I}} \\ \phi_{Q1} &= \frac{\pi}{2}\frac{\left[(V_{\pi Q} + \Delta V_{bQ}) + V_{\mathrm{pp}Q}\sin(2\pi f_m t + \Delta\theta)\right]}{V_{\pi Q}} \\ \phi_{Q2} &= \frac{\pi}{2}\frac{\left[(V_{\pi Q} + \Delta V_{bQ}) + (1+\Delta V_Q)V_{\mathrm{pp}Q}\sin(2\pi f_m t + \Delta\theta)\right]}{V_{\pi Q}} \\ \phi_{\mathrm{PM}} &= \frac{\pi}{2}\frac{(V_{\pi\mathrm{PM}} + \Delta V_{b\mathrm{PM}})}{V_{\pi\mathrm{PM}}} \end{aligned} \tag{3.6}$$

where ϕ_{RT} is the phase delay per RT. The phase difference induced by the phase modulators (PMs) for different arms are shown as follows: where $V_{\pi I}$, $V_{\pi Q}$ and $V_{\pi PM}$ represent the half-wave voltages of Sub-MZM1, Sub-MZM2, and PM respectively. ΔV_{bI}, ΔV_{bQ}, and ΔV_{bPM} represent the deviation DC bias voltage from the right bias operation point; ΔV_M (denoted as $\Delta V_M = V_{ppI} - V_{ppQ}$) and $\Delta\theta$ (denotes the phase shift between two RF deviating from $\pi/2$) are the amplitude imbalance and phase misalignment of the RF clock drive signals between I and Q ports. In (3.5) and (3.6), we can see that there are so many impact factors affecting the properties of I/Q modulator. Therefore, to achieve a good multitone output from I/Q modulator, we should apply the parameters properly in practice.

First, the imperfection of the I/Q modulator is studied in theory, including three direct current (DC) biases and the symmetry between A port and B port of I/Q modulator. Under this condition, we obtain the transfer function as follows:

$$\begin{aligned} T \approx \{k_0 + [k_{11}\cos(2\pi f_m t)) \\ +jk_{12}\sin(2\pi f_m t)] + k_2\cos(4\pi f_m t) \\ -[k_{31}\cos(6\pi f_m t) - jk_{32}\sin(6\pi f_m t)]\}\exp(j\phi_{\mathrm{RT}}) \end{aligned} \tag{3.7}$$

where the coefficients of the zeroth-order to third-order harmonics are as follows:

$$\begin{aligned} k_0 &= \frac{1}{2}[\sin(\Delta\delta_{bI})J_0(\delta_{mI}) + \mathrm{j}\exp(j\Delta\delta_{\mathrm{PM}})\sin(\Delta\delta_{bQ})J_0(\delta_{mQ})] \\ k_{11} &= \cos(\Delta\delta_{bI})J_1(\delta_{mI}) \\ k_{12} &= \exp(j\Delta\delta_{\mathrm{PM}})\cos(\Delta\delta_{bQ})J_1(\delta_{mQ}) \\ k_2 &= \exp(j\Delta\delta_{\mathrm{PM}})\sin(\Delta\delta_{bQ})J_2(\delta_{mQ}) - \sin(\Delta\delta_{bI})J_2(\delta_{mI}) \\ k_{31} &= \cos(\Delta\delta_{bI})J_3(\delta_{mI}) \\ k_{32} &= \exp(j\Delta\delta_{\mathrm{PM}})\cos(\Delta\delta_{bQ})J_3(\delta_{mQ}) \end{aligned} \tag{3.8}$$

Where $\Delta\delta_{bI} = (\pi\Delta V_{bI})/(2V_{\pi I})$, $\Delta\delta_{bQ} = (\pi\Delta V_{bQ})/(2V_{\pi Q})$, and $\Delta\delta_{PM} = (\pi\Delta V_{bPM})/(2V_{\pi PM})$ represent the bias voltage deviation from the right operation voltage respectively.

$\Delta\delta_{mI} = (\pi\Delta V_{ppI})/(2V_{\pi I})$ and $\Delta\delta_{mQ} = (\pi\Delta V_{ppQ})/(2V_{\pi Q})$ denote the deviation phase modulation due to intrinsic insertion of I/Q modulator. From (3.7) and (3.8), we can see that the even-order harmonics existed as long as the bias voltage deviation is not equal to zero. Obviously, to reduce the harmonics crosstalk, we should apply the bias voltage at the right operation voltage as accurate as possible in practice.

Besides, A port and B port of the I/Q modulator are not precisely symmetrical, therefore the null points of the two ports are not equal, which results in the zero-order harmonics cannot be completely converted to the desired signal.

Second, the influence of the unequal power and phase deviation from $\pi/2$ between two RF drive signals on the multitone generation is studied theoretically. Assuming that the I/Q modulator has no intrinsic insertion, loss and DC ports are operating at its right voltage, and the transfer function is expressed as follows:

$$\begin{aligned} T \approx \{[J_1(\delta_m)\exp(2\pi f_m t) - J_3(\delta_m)\exp(-6\pi f_m t)] \\ +j[k_{11}\sin(2\pi f_m t) + k_{12}\cos(2\pi f_m t)] \\ +j[k_{31}\sin(6\pi f_m t) + k_{32}\cos(6\pi f_m t)]\}\exp(j\phi_{\mathrm{RT}}) \end{aligned} \tag{3.9}$$

where the coefficients of the first-order and third-order harmonics are as follows:

$$\begin{aligned} k_{11} &= J_1(C_1)J_0(C_2) + J_1(C_1)J_2(C_2) - J_3(C_1)J_2(C_2) - J_1(\delta_m) \\ k_{12} &= J_1(C_2)J_0(C_1) + J_1(C_2)J_2(C_1) - J_3(C_2)J_2(C_1) \end{aligned}$$

$$\begin{aligned} k_{31} &= J_3(C_1)J_0(C_2) - J_1(C_1)J_2(C_2) - J_3(\delta_m) \\ k_{32} &= J_1(C_2)J_2(C_1) - J_3(C_2)J_0(C_1) \end{aligned} \tag{3.10}$$

where $\delta_m = \delta_{mI} = \delta_{mQ}$, $C_1 = (\delta_m + \Delta\delta_m)\cos(\Delta\theta)$, and $C_2 = (\delta_m + \Delta\delta_m)\sin(\Delta\theta)$.

Obviously, from (3.9) and (3.10), we can see that there are odd order harmonics crosstalk when the RF drive signals are not perfectly balanced.

Based on the previous analysis, we can draw a brief conclusion that: (1) the balance perfection of the I/Q modulator is very important. The deviation from right bias voltage will also generate the undesired even-order harmonics that can affect the output spectrum severely; (2) the power and phase imbalance of RF drive signals will generate the undesired odd-order tones. Both of the two cases will generate the undesired harmonics corresponding to the desired tone. Therefore, in the experiment, we should reduce the influence of imbalance factors to decrease the crosstalk components with appropriate operation.

3.1.2 *The Novel Double Sideband RFS Structure Based on DP-MZM*

Figure 3.4 shows the structure of multi-carrier generation based on DP-MZM, which comprises of CW laser, optical coupler, the DP-MZM modulator, two optical band pass filters, two EDFAs, and one PC. The continuous wave laser with the frequency of f0 is divided into two parts by an optical coupler (OC), one-part couples out and the other one enters the loop of the frequency shifting structure with two Mach–

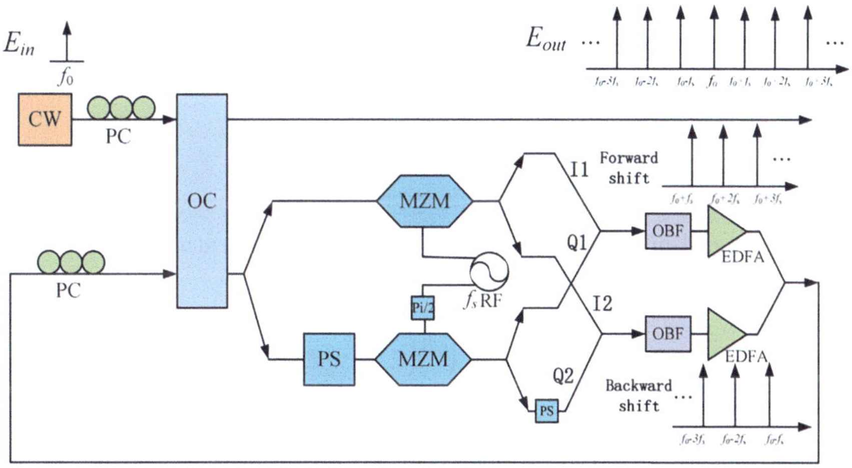

Fig. 3.4 Multi-carrier structure based on DPMZM (Dual Parallel Mach–Zehnder modulator), *CW* continuous wave, *PC* polarization controller, *OC* optical coupler, *PS* phase shifter, *MZM* Mach–Zehnder modulator, *OBF* optical bandpass filter, *EDFA* erbium doped fiber amplifier

Zehnder modulators (MZM) in parallel configuration. PS is used to generate 90-degree phase shift. The EDFA is used to compensate the frequency conversion loss. Two tunable bandpass filters with flat-top shape are used to select the number of carriers required, and the manual polarization controller is applied to realize the adaptive SOP control and stabilization of signals. The MZ modulator operates in the linear region, and the DC bias of the modulator is $V_\pi/2$, where V_π is the half-wave voltage of the modulator. The RF drive signals for two MZMs have the same amplitude and frequency with 90-degree phase shift. Assuming the RF signals applied to the two MZMs $S_1(t) = V_{pp}\cos(2\pi f_s t)$ and $S_2(t) = V_{pp}\sin(2\pi f_s t)$, respectively, the total voltage applied to the upper branch MZM is:

$$u_1(t) = S_1(t) + \frac{V_\pi}{2} = V_{pp}\cos(2\pi f_s t) \tag{3.11}$$

Since the modulator operates in the push–pull mode, the phase shift caused by the applied voltage is:

$$\varphi_1(t) = -\varphi_2(t) = \frac{u_1(t)}{V_\pi}\pi \tag{3.12}$$

and the output of the upper channel is:

$$E_1 = E_{cw}\cdot\frac{1}{2}\cdot\left(e^{j\varphi_1(t)} + e^{j\varphi_2(t)}\right) \tag{3.13}$$

Similarly, the voltage applied to the lower branch MZM is:

$$u_2(t) = S_2(t) + \frac{V_\pi}{2} = V_{pp}\sin(2\pi f_s t) \tag{3.14}$$

The phase shift between two arms of the lower branch MZM is:

$$\varphi_3(t) = -\varphi_4(t) = \frac{u_2(t)}{V_\pi}\pi \tag{3.15}$$

The lower signal first passes through 90°phase shifter and enters MZM, and the downstream output can be given as:

$$E_3 = E_{cw}\cdot\frac{1}{2}\cdot\left(e^{j\frac{\pi}{2}}\left(e^{j\varphi_3(t)} + e^{j\varphi_4(t)}\right)\right) \tag{3.16}$$

The output spectrum of the upper and lower branches can be expressed as (see Fig. 3.5):

The upper and lower signals respectively passed through 3 dB Y type waveguide, and are equally divided into two paths. The upper signal is divided into I1, I2 paths, and the down signal is equally divided into Q1, Q2 paths, where signals in path Q2

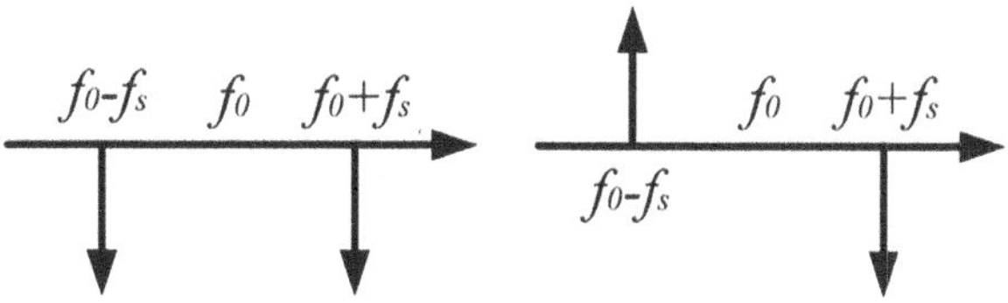

Fig. 3.5 The output spectrum of the upper and lower branches

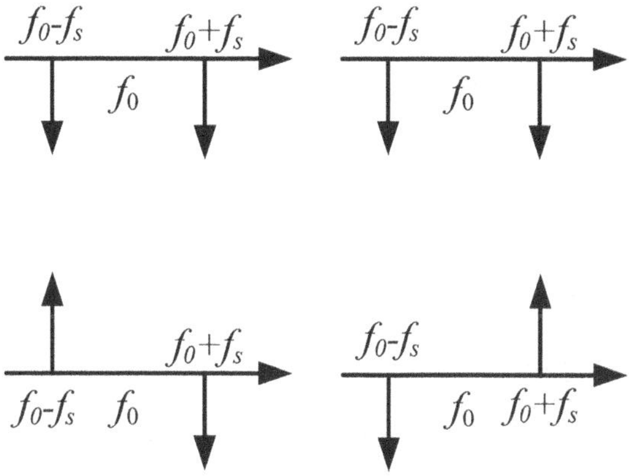

Fig. 3.6 The spectrum of I1, I2, Q1, Q2 signals

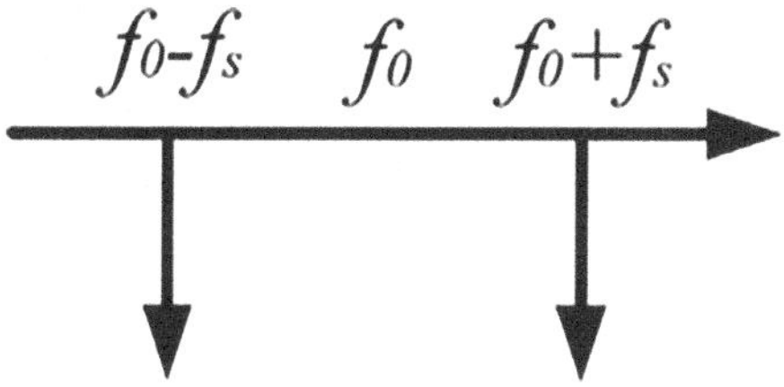

Fig. 3.7 The output after one loop

pass through the phase shifter which implements the flipping of the spectrum. The spectrum of I1, I2, Q1, Q2 signals can be expressed as (see Fig. 3.6):

The output signals I1 and Q1 are coupled as a forward frequency shifter, and the signals I2, Q2 are coupled as a backward frequency shifter. The forward frequency shifting signal passes through the optical bandpass filter, and the signal with the frequency greater than or equal to$f_0 + f_s$ is reserved. The reversed frequency shifted signal passes through the optical band pass filter, and the signal with the frequency less than or equal to $f_0 - f_s$ is reserved. The filters control the number of carriers. After the two signals are filtered, they are compensated by the EDFA. The spectrum can be shown as follows (see Fig. 3.7):

It can be seen that the input signal with the frequency f_0 passes through the loop and generates a signal with a frequency of $f_0 - f_s$ and $f_0 + f_s$, which realizes bidirectional frequency shift. After two-time loops, the frequencies $f_0 - f_s$ and $f_0 + f_s$ passes through the loop and generate signals with frequencies $f_0 - 2f_s$ and $f_0 + 2f_s$, realizing bidirectional frequency shifting. After N time loops, 2 N multi-carriers with a frequency range of $f_0 - Nf_s$ to $f_0 + Nf_s$ will be generated with the frequency interval of f_s. The optical bandpass filter is used to control the number of carriers according to the requirement.

3.2 Simulation and Experimental Results

3.2.1 Experiment on the Generation of 50 Stable Frequency-locked Optical Carriers Using an RFS

As 100 Gb/s ethernet has become increasingly a commercial reality, the next logical pressing issue is the migration path toward 1-Tb/s ethernet transportation. Recently, optical orthogonal frequency division multiplexing (O-OFDM) has been proposed and has become a promising technique for high spectral efficiency in the optical communication transmission [22–26]. To produce terabit per second (Tb/s) multiband coherent O-OFDM signal, the novel approach of multitone generation by employing a recirculating frequency shifter (RFS) architecture has been demonstrated to generate 24-tone and 36-tone in [27–29] respectively, which all obtained a Tb/s high data transmission rate. Although they do really perfect work on experiments and theories, the instability and unflatness of single-side band (SSB)-based RFS due to the operation factors that seriously affect the output performance have not been reported [30–33].

We have successfully generated 50 stable frequency-locked carriers. The operating principle of multicarrier generation is analyzed and some important factors that affect the stability of the multicarrier are experimentally investigated. These impact factors include the imbalance performance of the in phase/quadrature (I/Q) modulator, the accuracy of phase difference and the equality of the powers between two RF signals, the enough gain of the optical amplifier, the bandwidth and sharp window spectrum of the tunable filter, and the stability of the optical signals' states of polarization (SOP).

In the experiment, the laser source runs at its center wave-length of 1549.9 nm (f_0 = 193.425 THz), with the power of 12 dBm, which is shown in Fig. 3.8. To observe the symmetry of the I/Q modulator, the transmission curves of DC A and DC B have been measured as shown in Fig. 3.13. In the experiment, the voltages of A port and B port are biased at its null points of transmission curves respectively. RF signals are two clock signals with equal power but with $\pi/2$ phase shift, which can be accurately obtained by adjusting the attenuators to change RF power amplitude and adjusting the time delay of signal generator to change the phase difference between RF signals. The insertion loss of the I/Q modulator and the filter are 13 and 6 dB respectively. So, we used a Raman amplifier with its adjustable saturated output power range from 27 to 46 dBm. The amplifier facilitates the gain to balance the losses in the loop. The tunable bandpass filter with flat-top shape is used to select the required window of bands, which means that its bandwidth decides the number of the tones we desired to generate.

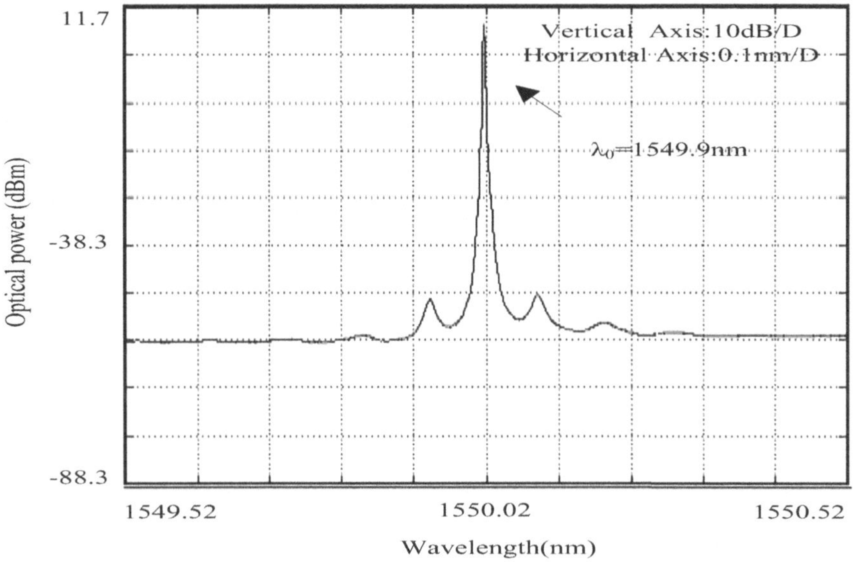

Fig. 3.8 Optical spectrum of the laser source

Fig. 3.9 Optical spectrum of the first tone generation when the loop is open

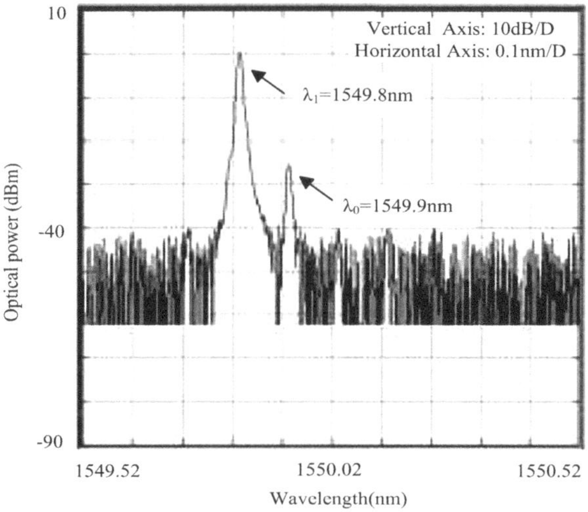

For the first round, only the first-order upward frequency ($f_1 = f_0 + f_m = 193.4375$ THz, *or* $\lambda_1 = 1549.8$ nm) is produced, as shown in Fig. 3.9. After the *N*th RT, the other tones such as $f_2, f_3, f_4, \cdots, f_N$ are generated, as shown in Fig. 3.10. The f_{N+1} tone and beyond are filtered out by the bandpass filter placed in the loop.

From Fig. 3.10, it can be seen that the generated stable 50 tone carriers are with superior flatness. In the experiment, we also found that several factors determine the

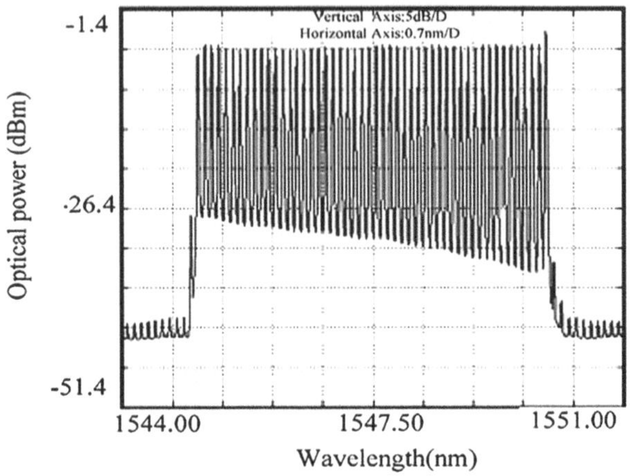

Fig. 3.10 Fifty stable frequency-locked-carrier spectrum

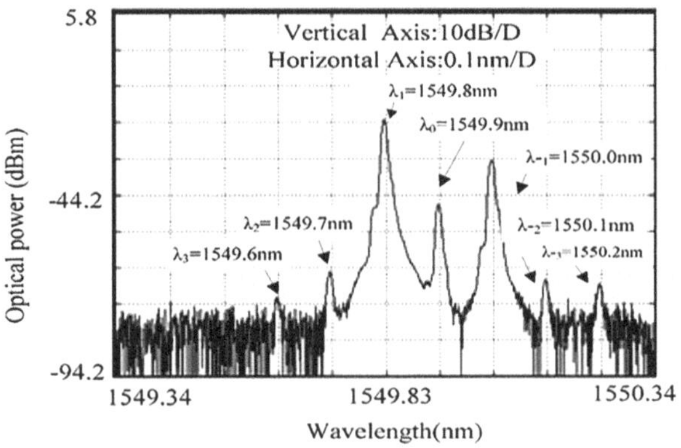

Fig. 3.11 Optical spectrum of the first tone generation with high-order harmonics

quality of tones, including the symmetry of I/Q modulator, the balance of two RF signals, the gain obtained from the optical amplifier, the bandwidth of the tunable filter, and the polarization controller.

Figure 3.9 shows the first tone generation when the influencing factors are balanced, the zero order cannot be completely converted to the desired signal (the first order) because two arms of the I/Q modulator are not precisely symmetrical, and the transmission curves of I/Q modulator have been shown in Fig. 3.3. The power difference of the null points (A port and B port) is round 0.08 mW. While Fig. 3.11 shows the first tone generation with many high-order harmonics when the DC biases voltage and RF signals are unbalanced. It is observed that DC biases voltage deviation mainly contributes to the even-order harmonics, both the amplitude and phase mismatch of RF signals significantly cause odd-order harmonics, inflicting interference on the desired tones. The results of these influences can be demonstrated by the comparison of Figs. 3.9 and 3.11.

In the experiment, we adopted two RF attenuators to adjust the voltage amplitude of RF signals and used the signal generator's delay time to change the phase

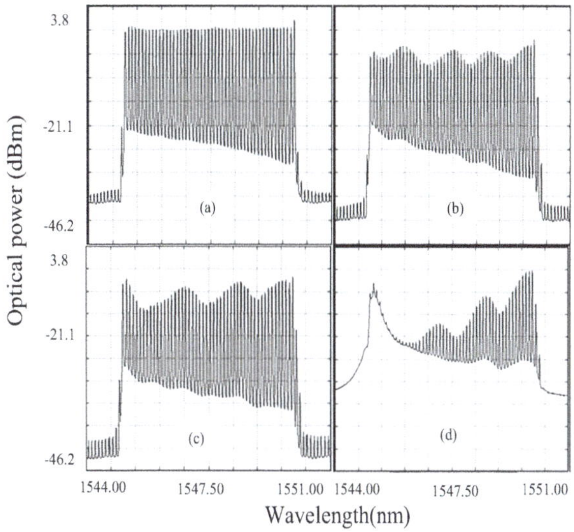

Fig. 3.12 Output spectrum with different power difference between two RF signals (**a**) 0 dB, (**b**) 0.5 dB, (**c**) 1 dB, and (**d**) 1.5 dB

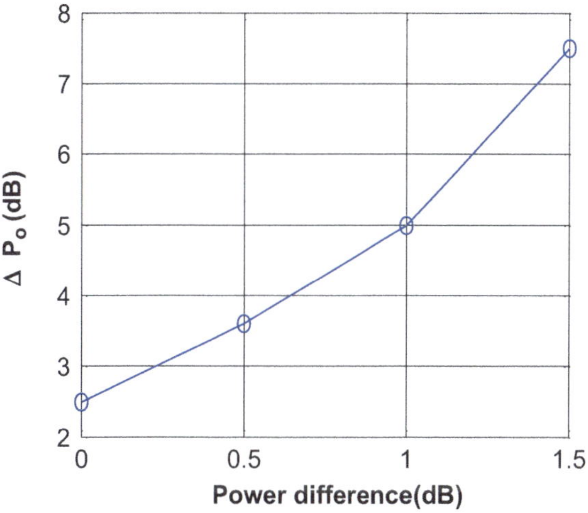

Fig. 3.13 Optical power difference of the highest tone to the lowest tone as a function of the power difference between two RF signals

difference of RF signals. Figures 3.12 and 3.13 show the results when the power difference of two RF clock signals is adjusted to 0, 0.5, 1, and 1.5 dB respectively. It can be seen that the flatness of the tones is degrading considerably as the power difference is increasing.

Figures 3.14 and 3.15 show the impact of the phase deviation $\Delta\theta$ (denotes the phase shift between two RF deviating from $\pi/2$) on the entire output of the multicarrier source. The parameters used in the figure are as follows: (a) $\Delta\theta = -\pi/2$; (b) $\Delta\theta = -\pi/4$; (c) $\Delta\theta = \pi/4$; (d) $\Delta\theta = \pi/2$. We can see that the output performance goes worse with the increase of $\Delta\theta$. It is confirmed that the accurate

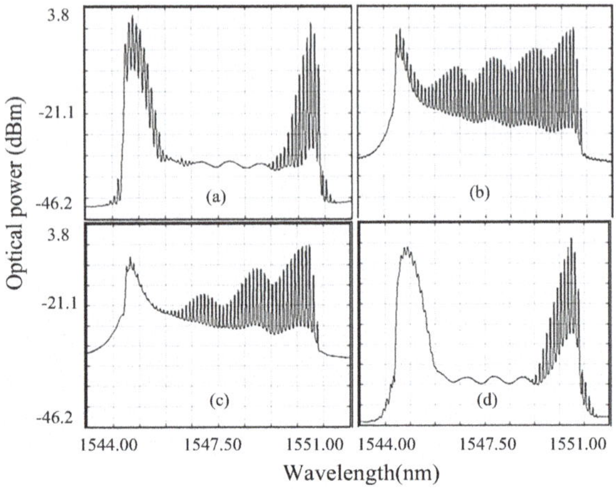

Fig. 3.14 Output spectrum with (**a**) $\Delta\theta = -\pi/2$, (**b**) $\Delta\theta = -\pi/4$, (**c**) $\Delta\theta = \pi/4$, and (**d**) $\Delta\theta = \pi/2$

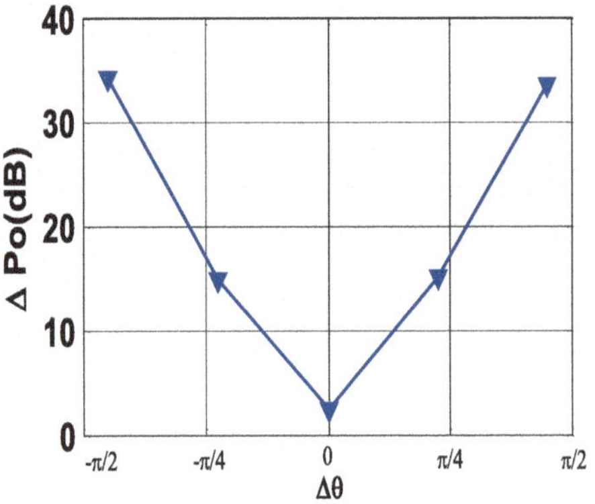

Fig. 3.15 Optical power difference of the highest tone to the lowest tone as a function of the phase deviation $\Delta\theta$

phase shift ($\Delta\theta = 0$) is very crucial for the stable multitone generation; if the phase shift is too much or too little, the optimal results cannot be obtained.

In order to investigate that how much the balance imperfection of the I/Q modulator affects the multicarrier generation, the deviation of A port DC bias voltage from the null point is set to 0.1, 0.3, 0.5, and 0.7 V, respectively. Figure 3.16 shows the variation of multi-tone spectrum, and Fig. 3.17 shows the power difference variation between the highest and the lowest tones with ΔV_{bias} equaling −0.7, −0.5, −0.3, −0.1, 0, 0.1, 0.3, 0.5 and 0.7 V respectively. It illustrates that the flatness is decreasing as the DC bias voltage is deviating from the null point.

The polarization controller is placed in the loop system in order to control the variation of the SOP and keep the output signal stable. Figure 3.18 shows an unstable case when the PC is moved out of the loop. In order to explain the role of PC, we designed an experiment that recorded the peak value of the tones every 2 s with a total record time of 30 min, either with PC or without PC. The results are shown in Fig. 3.19. The peak instability range is about 7 dB m without the PC in the loop, while only 0.5 dB m peak instability appears with PC in the loop. So, we found that

Fig. 3.16 Output spectrum with different DC bias voltage difference deviating from the null point (**a**) 0.1 V, (**b**) 0.3 V, (**c**) 0.5 V, and (**d**) 0.7 V

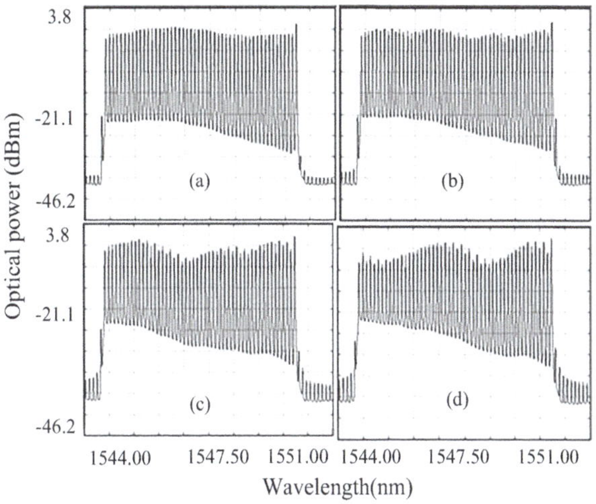

Fig. 3.17 Optical power diff. of highest to lowest tone vs. DC bias voltage diff. For A port

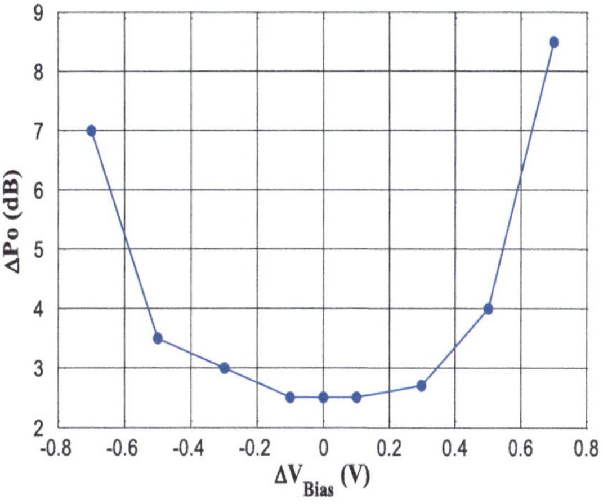

PC is one of the important parts of the RFS because the polarization stability plays the key role.

3.2.2 *Stability Analysis and Simulation Results of Multi-Carrier Source in Ideal Condition*

It is assumed that the optical amplifier can exactly compensate the modulation loss of the modulator and the device insertion loss in the link. Provided the polarization direction of the signal light is consistent with the modulation direction of the modulator, the fifth order can be ignored regardless of the filter. The crosstalk of the system's transfer function can be expressed as:

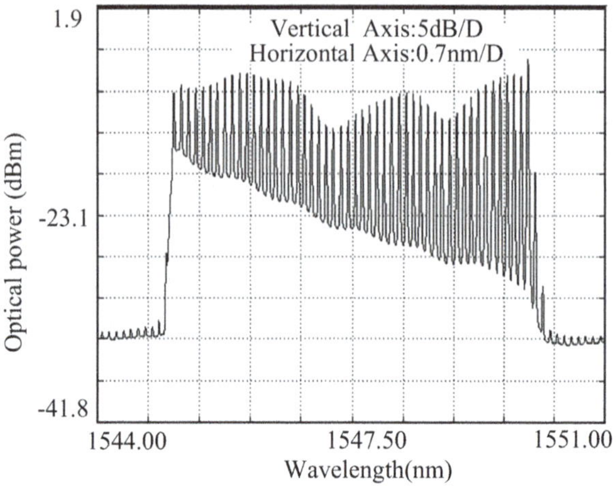

Fig. 3.18 Unstable 50 tone carriers spectrum

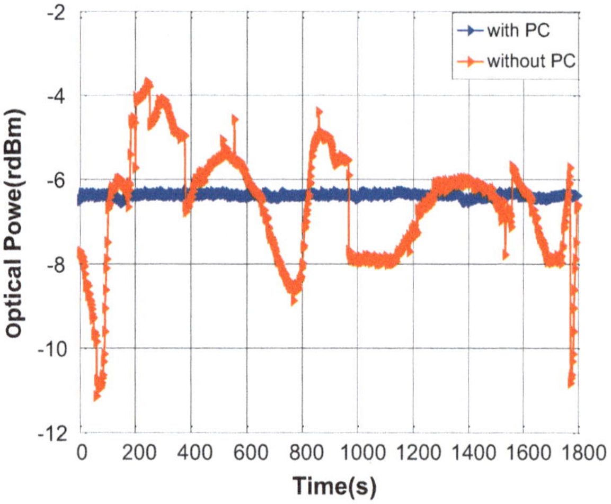

Fig. 3.19 Variation of the peak power with and without PC in the loop

$$T = \underbrace{\exp(j2\pi f_s t) + b\exp(-j6\pi f_s t)}_{forward} + \underbrace{\exp(-j2\pi f_s t) + b\exp(j6\pi f_s t)}_{backkwward} \quad (3.17)$$

where f_s is the frequency of the RF drive signal, and $b = -J_3(\delta_m)/J_1(\delta_m)$ is the normalized crosstalk coefficient. When considering the function of the filter, the forward shift retains a signal with a frequency larger than or equal to $f_0 + f_s$, and the reverse branch retains a signal with a frequency less than or equal to $f_0 - f_s$. The transfer function in this case can be expressed as:

$$T_1 = \underbrace{\exp(jw_s t)}_{forward} + \underbrace{\exp(-jw_s t)}_{backward} \tag{3.18}$$

where E represents the input optical signal, the output after first loop is $E \cdot T_1$, which contains the signal components of frequencies f_s and $-f_s$. The output of the second cycle can be expressed as:

$$\begin{aligned} E_2 = E \cdot & \underbrace{(\exp(j2w_s t) + \exp(-j2w_s t))}_{forward} + \underbrace{\exp(j \cdot 0 \cdot t) + \exp(j4w_s t))}_{backward} \\ + E \cdot & \underbrace{(\exp(j \cdot 0 \cdot t) + \exp(-j4w_s t))}_{forward} + \underbrace{\exp(-j2w_s t) + \exp(j2w_s t))}_{backutard} \end{aligned} \tag{3.19}$$

When considering the function of the filter, the transfer function of the second cycle can be expressed as:

$$T_2 = \underbrace{\exp(j2w_s t)}_{forward} + \underbrace{\exp(-j2w_s t)}_{backward} \tag{3.20}$$

Finally, the output of the second cycle can be expressed as $E_2 = E \cdot T_2$. It can be inferred that when entering the fifth cycle, the third-order crosstalk component will not be filtered. For the forward branch, the transfer function is

$$T_{forward} = \exp(j2\pi f_s t) + b\exp(-j6\pi f_s t) \tag{3.21}$$

For the backward shifter, the transfer function is:

$$T_{backward} = \exp(-j2\pi f_s t) + b\exp(j6\pi f_s t) \tag{3.22}$$

Therefore, the loop structure can be considered as two single side-band loop branches. The output of the fourth loop of signal components is given with frequencies $f_0 - 4f_s$ and $f_0 + 4f_s$. If the fourth output is $E_{-4} + E_4$, then the output of the fifth loop can be expressed as $E_5 = E_{-4}T_{backward} + E_4T_{forward}$. In order to minimize the influence of third-order crosstalk, the value of the normalized crosstalk coefficient b in the transfer function should be as small as possible. So, the multi-carrier output after the cycle of N times is:

$$\begin{aligned} E_N(t) &= E_0(t) + E_1(t) + E_2(t) + E_3(t) \\ &+ E_{-4}(t) \cdot \left(1 + T_{backward} + T_{backward}^2 + \cdots + T_{backward}^{N-4}\right) \\ &+ E_4(t) \cdot \left(1 + T_{forward} + T_{forward}^2 + \cdots + T_{forward}^{N-4}\right) \end{aligned} \tag{3.23}$$

where E_0 is the input signal, and $E_1(t)$, $E_2(t)$, $E_3(t)$ are the outputs after one, two and three cycles respectively. The above formula can be simplified as:

$$
\begin{aligned}
E_N(t) &= E_0(t) + E_1(t) + E_2(t) + E_3(t) \\
&+E_0(t)\cdot\sum_{n=4}^{N}\{\exp(j2\pi n f_s t) + \text{nbexp}(j2\pi(n-4)f_s t)\} \\
&+E_0(t)\cdot\sum_{n=4}^{N}\{\exp(-j2\pi n f_s t) + \text{nbexp}(-j2\pi(n-4)f_s t)\}
\end{aligned}
\tag{3.24}
$$

The output after N loop times:

$$
\begin{aligned}
E_{N+1}(t) &= E_N(t) + E_{-4}(t)\cdot T_{backward}^{N-3} + E_{forward}(t)\cdot T_{forward}^{N-3} \\
E_{N+2}(t) &= E_N(t) + E_{-4}(t)\cdot T_{backward}^{N-2} + E_{forward}(t)\cdot T_{forwar}^{N-2} \\
E_{N+3}(t) &= E_N(t) + E_{-4}(t)\cdot T_{backward}^{N-1} + E_{forward}(t)\cdot T_{forward}^{N-1} \\
E_{N+4}(t) &= E_N(t) + E_{-4}(t)\cdot T_{backward}^{N} + E_{forward}(t)\cdot T_{forward}^{N} \\
E_{N+5}(t) &= E_N(t) + E_{-4}(t)\cdot T_{backward}^{N+1} + E_{forward}(t)\cdot T_{forward}^{N+1} \\
E_{N+6}(t) &= E_N(t) + E_{-4}(t)\cdot T_{backward}^{N+2} + E_{forward}(t)\cdot T_{forward}^{N+2}
\end{aligned}
\tag{3.25}
$$

It can be seen that, $E_{N+4}(t)$, $E_{N+5}(t)$, $E_{N+5}(t)$ are approximately equal, the noise component is basically stabilized, and the entire system reaches to a steady state.

The stability of the dual-band multi-carrier based on a dual Mach–Zehnder modulator with its theoretical derivation and validation have been demonstrated in Fig. 3.20. The spectrum of the multi-carrier is simulated based on previous theoretical analysis. It shows the variation of the number of carriers and flatness with the RF drive voltage reflects the influence of the crosstalk component on the signal component. To obtain the desired tone number, the RF drive voltage is $0.18V_\pi$ with the best flatness.

Figure 3.20 presents the flatness of multi-carrier with different V_{pp}. When $V_{pp} = 0.18V_\pi$, and the flatness of the structure is perfect. In the simulation, the RF drive voltage is set $V_{pp} = 0.18V_\pi$, the frequency of the RF drive signal is $f_s = 1.25 \times 10^{10}$ Hz, the parameters of the two Mach–Zehnder modulators are the same, the half-wave voltage is $V_\pi = 4V$, and the azimuth of the polarization controller ensures the polarization of the polarized light. The filter is an ideal band-pass filter, and the optical amplifier is used to compensate the loss in the link.

From Fig. 3.21 it can be seen that when looping 30 times, the double-sided recirculating frequency shift generates a total of 60 carriers with a frequency interval of $f_s = 1.25 \times 10^{10}$ Hz, and the flatness is 0.6026 dB. When the cycle is 50 times, 100 carriers are generated with a frequency interval of $f_s = 1.25 \times 10^{10}$ Hz with flatness of 0.7769 dB. Similarly, when the cycle is 100 times, the double-side cyclic frequency shift generates a total of 200 carriers with a frequency interval of $f_s = 1.25 \times 10^{10}$ Hz with flatness of 1.148 dB. In general, the flatness decreases with the increase in the number of multi-carriers.

Figure 3.22 presents the spectrum compared of the single side band (SSB) recirculating frequency shifter with the proposed double side band structure after looping 30 times. Moreover, we have supplied the comparison of flatness against

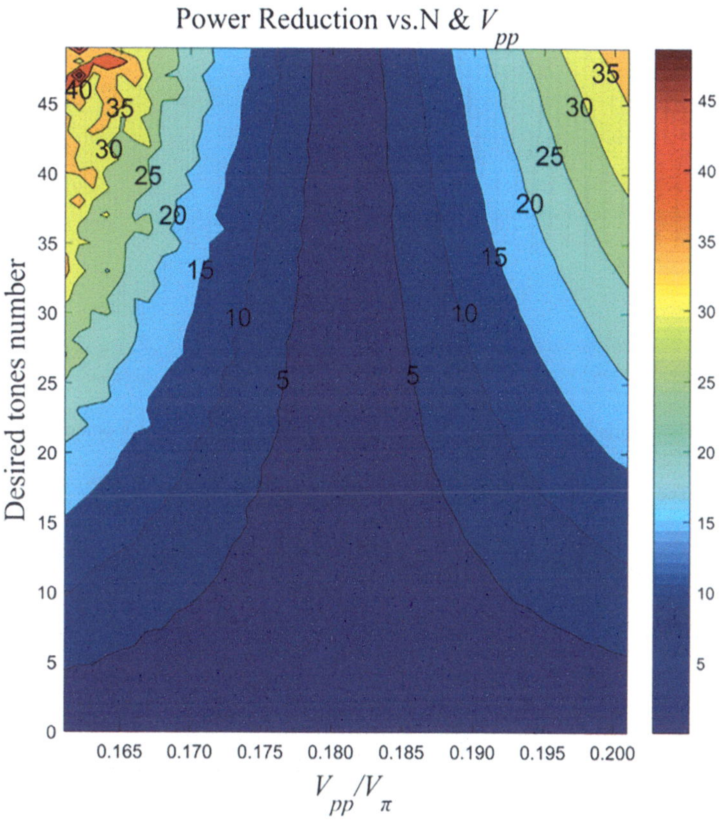

Fig. 3.20 The flatness with different V_{pp}

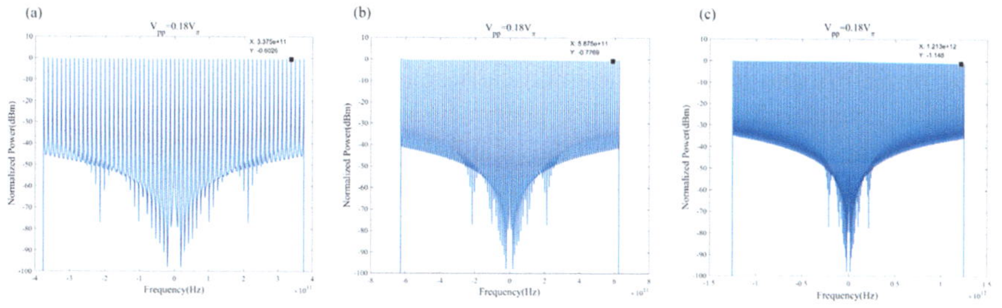

Fig. 3.21 The simulation results that when looping 30 times, 50 times, 100 times

different loop times for the mentioned two schemes shown in Table 3.1, which shows that the proposed scheme has better flatness and the number of carriers doubled with the same number of cycles.

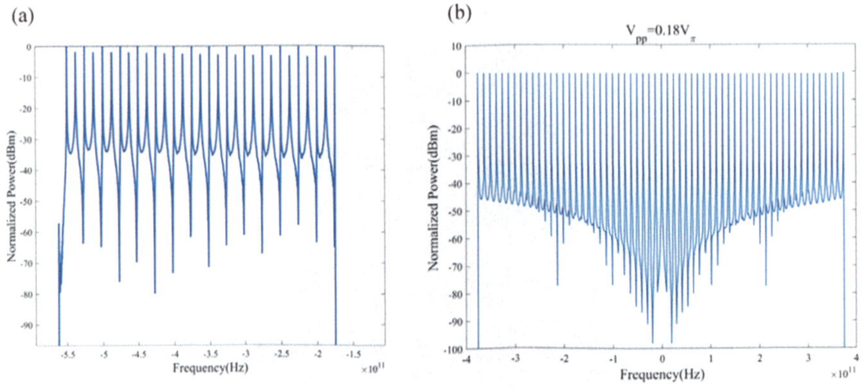

Fig. 3.22 The simulation results of SSB-RFS (**a**) and DSB-RFS (**b**) when looping 30 times

Table 3.1 The corresponding coordinates of 8 spaces

Loop number		10	20	30	40	50
Number of carriers	SSB-RFS	10	20	30	40	50
	DSB-RFS	20	40	60	80	100
Flatness (dB)	SSB-RFS	2.704	3.196	3.502	3.603	3.724
	DSB-RFS	0.405	0.418	0.431	0.523	0.777

3.2.3 *Stability Analysis and Simulation Results of Multi-carrier Source with ASE Noise of EDFA*

In the structure of multi-carrier generation based on DP-MZM, the EDFA is used to compensate the frequency conversion loss, which also introduce the amplified spontaneous emission (ASE) noise. The flatness of carriers is strongly influenced by the noise figure of EDFA. Therefore, investigating the properties of the ASE noise of EDFA that influence the multi-carriers is important. When ASE noise of EDFA is considered, the ASE noise generated at the output can be expressed by $P_{ASE_{total}}$ in the case of N cycles, and the total noise can be expressed as:

$$
\begin{aligned}
P_{ASE_{total}} = P_{ASE1} \cdot L_1 \cdot H_2 \cdot \ldots \cdot H_{N-1} \cdot G_N + \\
P_{ASE2} \cdot L_2 \cdot H_3 \cdot \ldots \cdot H_{N-1} \cdot G_N + \\
P_{ASE(n-1)} \cdot L_{n-1} \cdot G_N + P_{ASEn}
\end{aligned}
\tag{3.26}
$$

where n is the number of cycles, G is the gain of EDFA, L is the loop loss, and ASE is the noise introduced by EDFA. HN represents the product between the loss of the loop and gain of EDFA in the Nth cycle. The noise power in each cycle including two carriers can be expressed as:

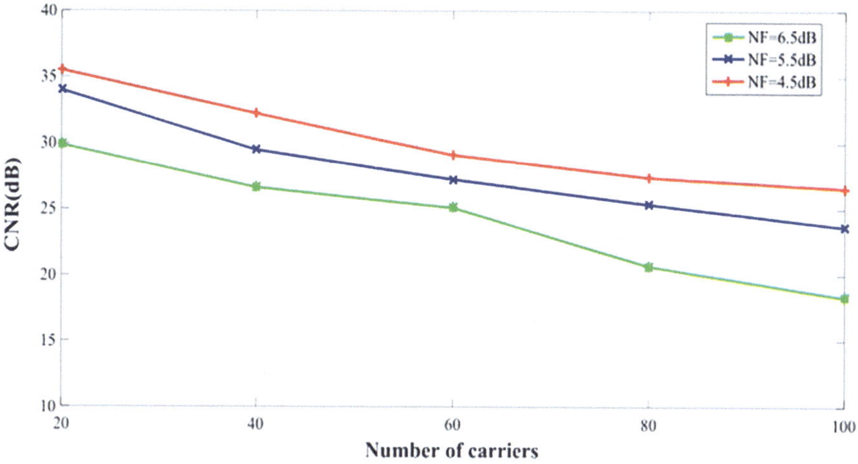

Fig. 3.23 The relationship between the number of carriers and CNR with different NF

$$P_{ASEn} = 2\left(NF_n - \frac{1}{G_n}\right)hvB_0G_n \tag{3.27}$$

The total gain in the system is:

$$G_{total} = H_1 \cdot H_2 \cdot H_3 \cdot \ldots \cdot H_{N-1} \cdot G_N \tag{3.28}$$

If the loss of the fiber loop is equal to the gain of EDFA, H = 1, then the equivalent noise figure of a cascaded model in the proposed system can be expressed as:

$$\begin{aligned} NF &= \frac{P_{ASE_{total}}}{2hvB_0G_{total}} + \frac{1}{G_{total}} \\ &= NF_1 + \frac{NF_2 - \frac{1}{G_2}}{G_1} + \frac{NF_3 - \frac{1}{G_3}}{G_1 \cdot G_2} + \cdots + \frac{NF_N - \frac{1}{G_N}}{G_1G_2\cdots G_{N-1}} \end{aligned} \tag{3.29}$$

Among them, NF_N denotes the noise figure of the Nth cycle EDFA, GN denotes the gain of EDFA in the Nth cycle, and NF denotes the total noise figure. For the *i*th carrier output, the noise power can be expressed as:

$$P_{ASE}(i)_{dBm} = 10 \cdot \log_{10}[i \cdot P_{ASE1}] = 10 \cdot \log_{10}[i \cdot (NF \cdot G - 1) \cdot hv \cdot B_0] \tag{3.30}$$

where hv denotes photon energy, B_0 presents the bandwidth of each carrier. It can be seen that the noise figure of EDFA determines the noise power of subcarriers. We numerically simulate the effects of EDFA noise figure on the output carrier-to-noise ratio (CNR), as shown in Fig. 3.23.

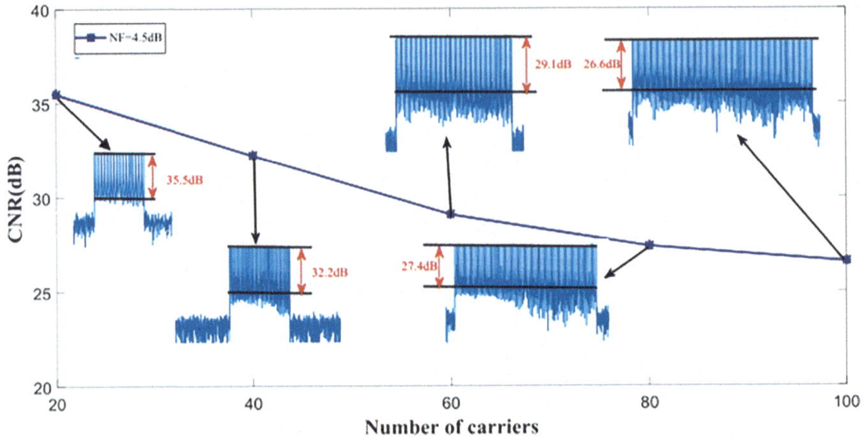

Fig. 3.24 The CNR variation with different number of carriers when $NF = 4.5\ dB$

From Fig. 3.20, it can be seen the CNR of the carriers is degrading with the increase of noise figure, and the CNR decreases quickly with a greater number of output carriers. In the following simulation, we perform the noise figure at 4.5. The output of the carriers is shown in Fig. 3.24. It can be seen that the amount of noise corresponding to each case of carriers' is superimposed with cycles, and the increasing noise is reflected in the spectrum.

Based on the above analysis of ASE noise for EDFA, the desired carrier's number is 2 N for the dual-loop configuration, and the stable output of multi-carrier after the Nth loop can be written as:

$$\begin{aligned} E_1 &= E_{in}(t) + \left[E_{in}(t)\cdot T_{forward}(t)\cdot g_1\cdot l \right.\\ &\quad \left. + E_{in}(t)\cdot T_{backward}(t)\cdot g_1\cdot l + n_{ASE}(t)\right]\\ E_2 &= E_{in}(t) + \left[E_1(t)\cdot T_{forward}(t)\cdot g_2\cdot l \right.\\ &\quad \left. + E_1(t)\cdot T_{backward}(t)\cdot g_2\cdot l + n_{ASE}(t)\right]\\ E_n &= E_{in}(t) + \left[E_{n-1}(t)\cdot T_{forward}(t)\cdot g_n\cdot l \right.\\ &\quad \left. + E_{n-1}(t)\cdot T_{backward}(t)\cdot g_n\cdot l + n_{ASE}(t)\right] \end{aligned} \tag{3.31}$$

where l and g_n represent the loss of the system optical fiber loop and the gain of EDFA, n_{ASE} represents the noise components in each cycle.

In order to analyze the influence of noise introduced by EDFA on the flatness of multi-carrier, Figure 3.25 shows the effect of ASE noise on the output CNR of carriers, in which red line demonstrates that the influence of ASE noise is considered, and blue line presents that without ASE noise. It can be seen that the influence of ASE noise increases and the CNR decreases with the increase of the number of cycles.

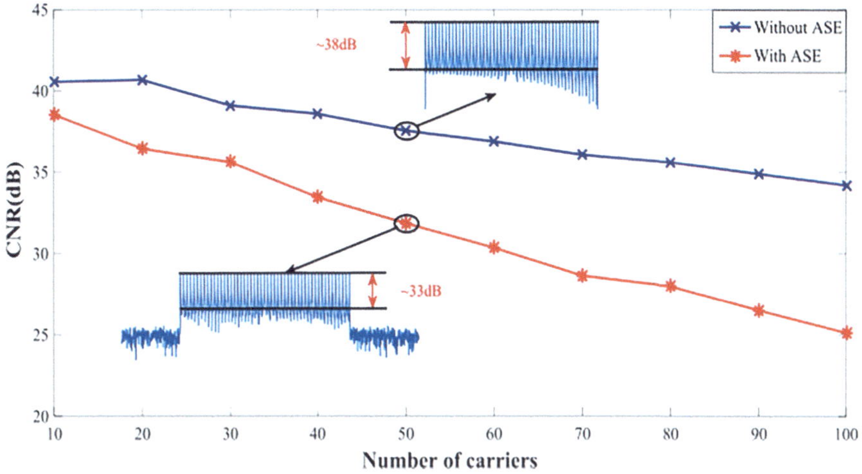

Fig. 3.25 The numerical results comparison w/wo consideration of ASE noise with EDFA. (For interpretation of the references to color in this figure legend, the reader is referred to the web version of this article)

3.3 RFS-Generated 20-Carrier Optical Comb for PDM-RZ-16QAM CO-WDM

In this subsection, we experimentally generate 20-tone, 25 GHz spacing carriers based on recirculating frequency shifter technology achieving the TNR of larger than 25 dB which could be used for 16 quadrature amplitude modulation (16QAM) transmission. Therefore, we modulate 16QAM signals at 16 Gbaud on each of the 25 GHz spacing subcarriers, and combine two branches of 16QAM into polarization division multiplexing-return zero-16 quadrature amplitude modulation (PDM-RZ-16QAM) format. Thereafter we realized totally a 2.56-Tb/s, PDM-RZ-16QAM, 25GHz spacing coherent optical wavelength division multiplexing (CO-WDM), 792 km transmission with coherent detection. Compared to the theoretical limits, the measured optical signal to noise ratio (OSNR) sensitivity curve is shifted by approximately 4 dB at a bit-error rate (BER) of 1e-3.

3.3.1 Experiment Setup and Transmission System Configuration

Figure 3.26 shows the experiment setup for the 2.56-Tb/s PDM-RZ-16QAM CO-WDM transmission system with the multi-carrier source. The wavelength of the light source (CW2) at 1543.06 nm is modulated by I/Q modulator with 16 Gb/s 4 level data which is generated by a pulse pattern generator. The subsequent return to

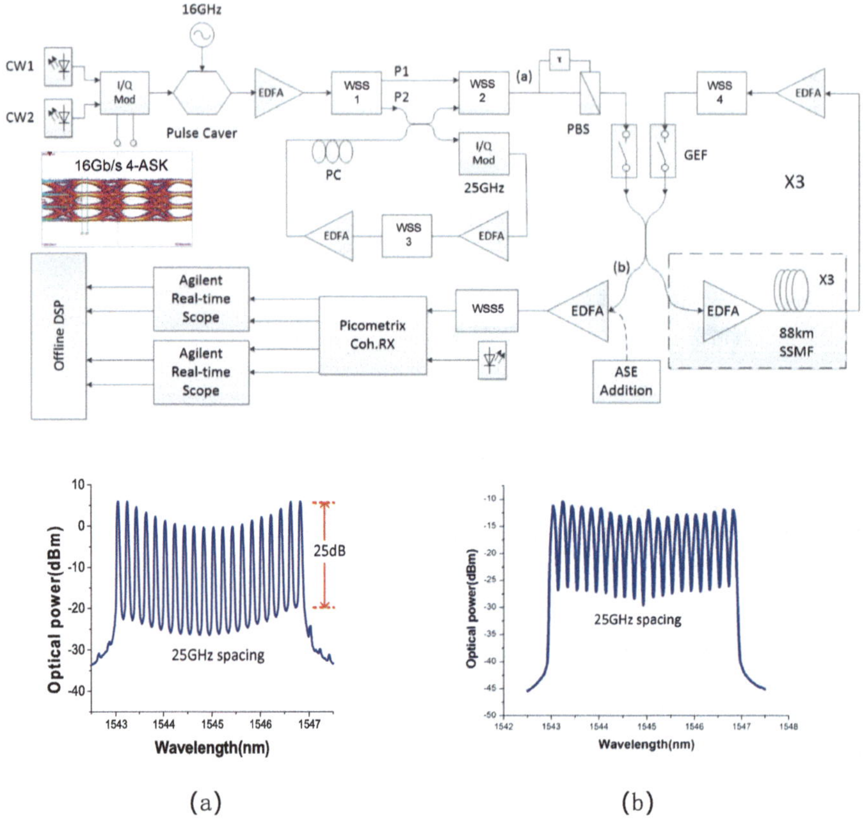

Fig. 3.26 Experiment setup for the 2.56-Tb/s 16-QAM CO-WDM transmission system (**a**) The spectrum of twenty carriers before transmission (**b**) The spectrum of twenty carriers after transmission

zero (RZ) carver helps to carve NRZ signal to RZ signal pulse. Then the signal outputs from the port2 of the WSS1 and then inputs a recirculating frequency shifter to generate twenty, 25 GHz-spaced tones in which each carrier carries 64-Gb/s 16QAM shown in Fig. 3.26a. The wavelength of the light source (CW1) at 1545.06 nm is used as the test channel, it carries RZ-16QAM signal and directly passes through WSS1 from P1 port to WSS2. In the WSS2, the CW1 signal replaces the central channel of the twenty carriers from the RFS. Finally, all of the channels are polarization-multiplexed to form a 20 CO-WDM channels PDM-RZ-16QAM signal with totally 2.56 Tb/s for 792 km transmission. The recirculating loop consists of three spans of 88 km of standard single-mode fiber, where EDFA is used to compensate the loss in the fiber. The loop is controlled with acousto-optic modulators (AOMs), which are drawn as switches in the Fig. 3.26. GEF means gain equalizing filter, it is used to ensure that all of the channels propagate through the recirculating loop with even power. WSS4 is used to select one of the channels for digitization and demodulation.

After transmission, the data is unloaded and detected with a Picometrix Coherent Receiver and digitized at 80 GS/s with two Agilent real-time oscilloscopes. Offline processing consists of (1) blind chromatic dispersion estimation and equalization, (2) timing phase recovery using the square-and-filter method, (3) polarization de-multiplexing with independent component analysis (ICA), (4) frequency offset estimation, (5) carrier phase recovery and (6) per-polarization LMS equalization.

For each channel we varied the launch power per-channel from −5 dBm to +4 dBm to cover a reasonable range of the power in commercial transmission. At each launch power we measured the BER vs. OSNR performance of the center 16QAM channel.

3.3.2 Experiment Results

The back-to-back OSNR requirements of 128 Gb/s (16 Gbaud) PDM-RZ-16QAM signals are shown in Fig. 3.27. The measurement was taken by using WSS4 to select one of the channels concerned and the bandwidth is set as 25 GHz. Compared to the theoretical limit the measured OSNR sensitivity curve for single channel is shifted by approximately 3.2 dB at a BER of 1e-3. The reason why for this penalty is: the theoretical limit is obtained only considering the ASE noise from EDFAs in the transmission line with VPI software, while in the experiment the optical signals would suffer from the other effects of residual uncompensated CD (chromatic dispersion), PMD (polarization mode dispersion), incomplete polarization de-multiplexing, and SPM as well for single channel. The OSNR sensitivity curve for WDM channel has another penalty of 0.9 dB compared to single channel, which is due to the XPM crosstalk from neighboring channels.

The constellations of one polarization of 16-Gbaud PDM-RZ-16QAM signals for single channel and 20-channel at bit-error rate (BER) of 1e-3 are shown in Fig. 3.28 a and b, respectively.

The 2.56 Tb/s CO-WDM PDM-RZ-16QAM was transmitted over 792 km fiber link. Figure 3.29 shows the dependence of the BER vs. the launch power

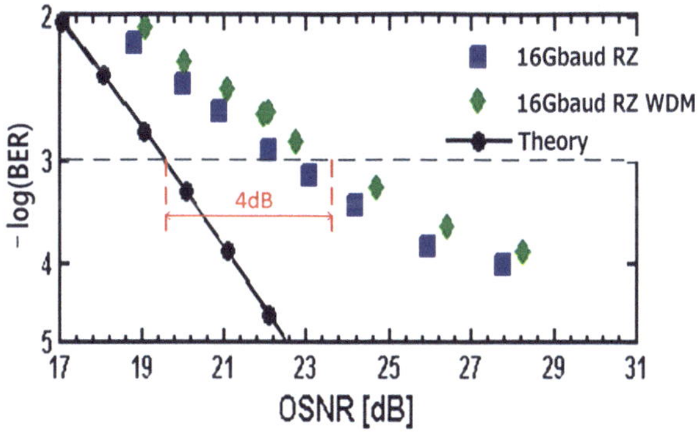

Fig. 3.27 Measured B2B OSNR requirement

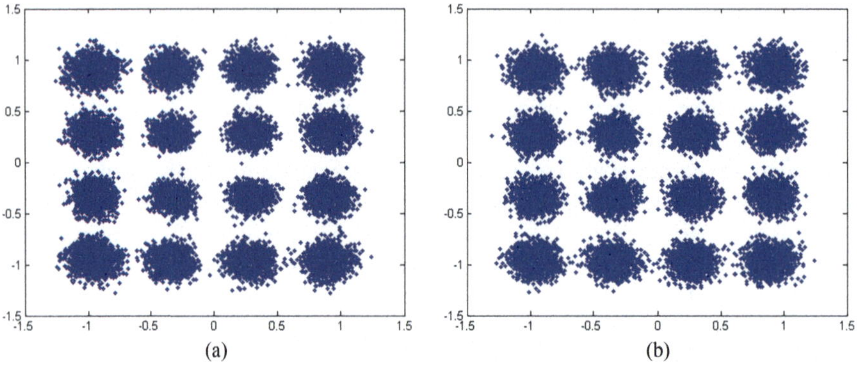

Fig. 3.28 PDM-RZ-16QAM constellation diagrams for (**a**) BTB signal channel x-branch data, (**b**) BTB central channel x-branch data of multi-channels

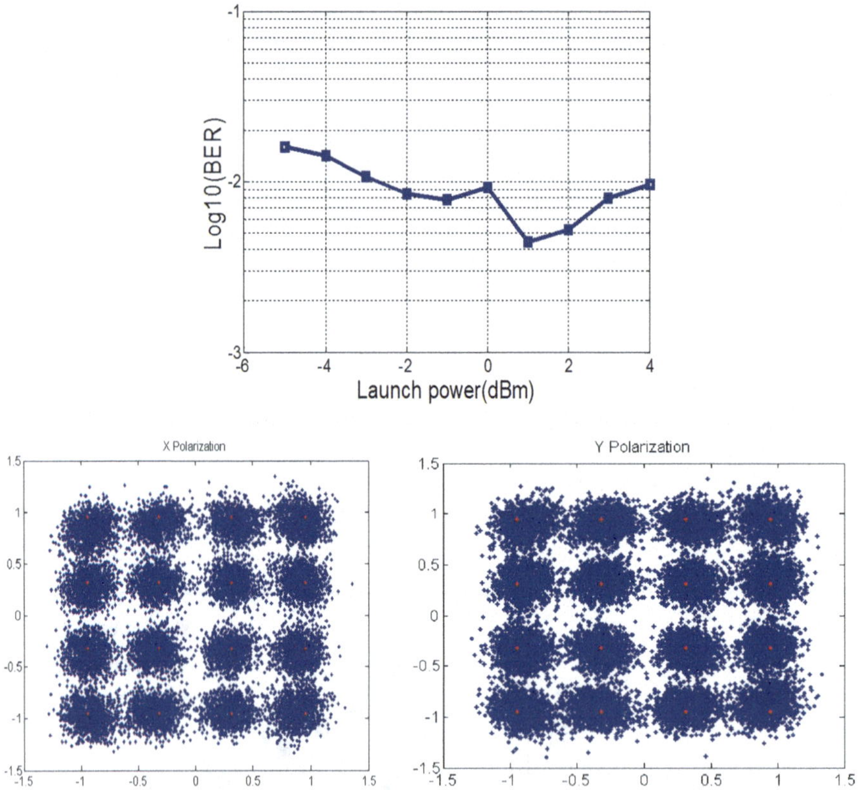

Fig. 3.29 (**a**) BER results vs. launch power for the central transmitted channel. (**b**) and (**c**) correspond to the constellation diagrams of x-branch and y-branch data

per-channel, indicating the optimal launch power for the system is 1 dBm. The BER value of 4.5e-3 at 1 dBm launch power is under the threshold value of soft-decision FEC 2.4e-2. The constellations of x and y polarizations for recovered CO-WDM PDM-RZ-16QAM signals chosen from one of channels at the optimal power (BER = 4.5e-3) is shown in Fig. 3.29. The spectral efficiency for our transmission system is 5.1 bit/s/Hz, which is better than traditional WDM system with the spectral efficiency 4.2 bit/s/Hz [34].

3.4 Summary

This chapter focus on the progress in multi-carrier generation and large capacity optical transmission systems via RFS technology. Initially, an SSB modulator—based RFS is utilized to produce 50 frequency—locked optical carriers with 12.5—GHz spacing, featuring remarkable flatness and stability. Key factors influencing carrier flatness, such as I/Q drive signal balance, time alignment, modulator bias accuracy, and polarization stability, are experimentally examined. Subsequently, a new dual—band RFS architecture based on a DP—MZM is put forward, which doubles efficiency and cuts costs in half. Simulations show that under optimized RF drive conditions, 60–200 carriers can be generated with flatness as low as 0.6 dB and carrier—to—noise ratios (CNR) over 40 dB. Finally, by using RFS the generated carriers (TNR > 25 dB), a 2.56—Tb/s PDM—RZ—16QAM CO—WDM system is experimentally verified over 800 km of standard single-mode fiber, attaining a spectral efficiency of 5.1 b/s/Hz.

References

1. A.D. Ellis, F.C. Gunning, B. Cuenot, T.C. Healy, E. Pincemin, Towards 1TbE using coherent WDM, in *OECC/ACOFT 2008-Joint Conference of the Opto-Electronics and Communications Conference and the Australian Conference on Optical Fibre Technology*, (IEEE, 2008), pp. 1–4
2. R. Dischler et al., Transmission of 1.2 Tb/s continuous waveband PDM-OFDM-FDM signal with spectral efficiency of 3.3 bit/s/Hz over 400 km of SSMF, in *National Fiber Optic Engineers Conference 2009 Mar 22*, (Optica Publishing Group, 2009), p. PDP2
3. R. Dischler, F. Buchali, Efficient digital coherent detection of a 1.2-Tb/s 24-carrier no-guard-interval CO-OFDM signal by simultaneously detecting multiple carriers per sampling, in *OFC'10*, (2010), p. OWO2
4. S. Chandrasekhar, X. Liu, Terabit superchannels for high spectral efficiency transmission, in *36th European Conference and Exhibition on Optical Communication*, (IEEE, 2010), pp. 1–6
5. X. Liu, S. Chandrasekhar, B. Zhu, D.W. Peckham, Transmission of a 1.2-Tb/s 24-carrier no-guard-interval coherent OFDM superchannel over 7200-km of ultra large-area fiber, in *ECOC'09, Post-Deadline Paper PD2.6*, (2009)
6. R. Ullah, L. Bo, S. Ullah, M. Yaya, F. Tian, X. Xiangjun, Cost effective OLT designed from optical frequency comb generator based EML for1. 22 Tbps wavelength division multiplexed passive optical network. Opt. Fiber Technol. Mater. Dev. Syst. **31**(43), 49–56 (2018)

7. R. Ullah, B. Liu, Q. Zhang, M.S. Khan, I. Ahmad, A. Ali, R. Khan, Q. Tian, C. Yan, X. Xin, Pulsed laser-based optical frequency comb generator for high-capacity wavelength division multiplexed passive optical network supporting 1.2 Tbps. Opt. Eng. **55**(9), 096106 (2016)
8. T. Healy, F.C. Garcia Gunning, A.D. Ellis, J.D. Bull, Multi-wavelength source using low drive-voltage amplitude modulators for optical communications. Opt. Express **15**, 2981–2986 (2007)
9. Y. Li, Z. Liu, J. Zhou, Y. Qiao, Y. Ji, Theoretical and experimental analysis of multi-carriers based on re-circulating frequency shifter, in *Asia Communications and Photonics Conference 2013, OSA Technical Digest*, (Optical Society of America, 2013) paper AF3F.4
10. F. Tian, X. Zhang, J. Li, L. Xi, Generation of 50 stable frequency-locked optical carriers for Tb/s multicarrier optical transmission using a recirculating frequency shifter. J. Lightwave Technol. **29**, 1085–1091 (2011)
11. J. Li, X. Li, X. Zhang, F. Tian, L. Xi, Analysis of the stability and optimizing operation of the single-side-band modulator based on re-circulating frequency shifter used for the t-bit/s optical communication transmission. Opt. Express **18**, 17597–17609 (2010)
12. H. Tu, L. Xi, X. Zhang, X. Zhang, J. Lin, W. Meng, Analysis of the performance of optical frequency comb based on recirculating frequency shifter influenced by an Er-doped fiber amplifier. Photon. Res. **1**, 88–91 (2013)
13. X. Li, J. Yu, Z. Dong, J. Zhang, Y. Shao, N. Chi, Multi-channel multi-carrier generation using multi-wavelength frequency shifting recirculating loop. Opt. Express **20**, 21833–21839 (2012)
14. J. Zhang, Y. Jianjun, Z. Dong, Y. Shao, N. Chi, Generation of full C-band coherent and frequency-lock multi-carriers by using recirculating frequency shifter loops based on phase modulator with external injection. Opt. Express **19**, 26370–26381 (2011)
15. J. Li, Z. Li, Frequency-locked multicarrier generator based on a complementary frequency shifter with double recirculating frequency-shifting loops. Opt. Lett. **38**(3), 359–361 (2013)
16. J. Li, C. Yu, Z. Li, Complementary frequency shifter based on polarization modulator used for generation of a high-quality frequency-locked multicarrier. Opt. Lett. **39**(6), 1513–1516 (2014)
17. J. Zhang, N. Chi, J. Yu, J. Zhu, Y. Shao, B. Huang, L. Tao, Generation of 113 coherent and frequency-lock multicarriers using cascaded phase modulators and EDFA loop for Tb/s optical communication, in *37th European Conference and Exposition on Optical Communications, OSA Technical Digest (CD)*, (Optical Society of America, 2011) paper We.10. P1.07
18. Y. Wang, N. Chi, J. Zhang, S. Zou, Investigation on the generation of coherent optical multi-carriers using cascaded phase modulators, in *Asia Communications & Photonics Conference & Exhibition*, (IEEE, 2011)
19. J. Yu, Z. Dong, J. Zhang, et al., Generation of coherent and frequency-locked multi-carriers using cascaded phase modulators for 10 Tb/s optical transmission system. J. Lightwave Technol. **30**(4), 458–465 (2012)
20. J. Yu, X. Li, J. Yu, N. Chi, Flattened optical frequency-locked multi-carrier generation by cascading one directly modulated laser and one phase modulator. Chin. Opt. Lett. **11**, 110606 (2013)
21. J. Li, Z. Li, C. Lu, A novel frequency-locked multicarrier generator by utilizing the three-branch-waveguide-interferometer-based single-side-band modulator, in *Advanced Photonics for Communications, OSA Technical Digest*, (Optical Society of America, 2014) paper JT3A.8
22. W. Shieh, C. Athaudage, Coherent optical orthogonal frequency division multiplexing. Electron. Lett. **42**(10), 587–589 (2006)
23. W. Shieh, X. Yi, Y. Tang, Transmission experiment of multi-gigabit coherent optical OFDM systems over 1000 km SSMF fiber. Electron. Lett. **43**(3), 183–185 (2007)
24. S.L. Jansen, I. Morita, N. Takeda, H. Tanaka, 20-Gb/s OFDM transmission over 4 160-km SSMF enabled by RF-pilot tone phase noise compensation, in *Presented at the Optical Fiber Communication Conference, Anaheim, CA*, (2007) Paper PDP15
25. S. Chandrasekhar, X. Liu, Experimental investigation on the performance of closely spaced multi-carrier PDM-QPSK with digital coherent detection. Opt. Exp. **17**, 21350–21361 (2009)
26. W. Shieh, Q. Yang, Y. Ma, 107 Gb/s coherent optical OFDM transmission over 1000-km SSMF fiber using orthogonal band multiplexing. Opt. Exp. **16**, 6378–6386 (2008)

27. X. Liu, S. Chandrasekhar, B.Y. Zhu, D.W. Peckham, Efficient digital coherent detection of a 1.2-Tb/s 24-carrier no-guard-interval CO-OFDM signal by simultaneously detecting multiple carriers per sampling, in *Proceedings of the Optical Fiber Communication (OFC), National Fiber Optic Engineers Conference (NFOEC)*, (2010), pp. 1–3
28. S. Chandrasekhar, X. Liu, B. Zhu, D.W. Peckham, Transmission of a 1.2-Tb/s 24-carrier no-guard-interval coherent OFDM super-channel over 7200-km of ultra-large-area fiber, in *Proceedings of the European Conference on Optical Communications (ECOC)*, (2009), pp. 1–2
29. Y. Ma, Q. Yang, Y. Tang, S.M. Chen, W. Shieh, 1-Tb/s single-channel coherent optical OFDM transmission with orthogonal-band multiplexing and subwavelength bandwidth access. J. Lightwave Technol. **28**(4), 308–315 (2010)
30. X. Liu, D.M. Gill, S. Chandrasekhar, L.L. Buhl, M. Earnshaw, M.A. Cappuzzo, L.T. Gomez, Y. Chen, F.P. Klemens, E.C. Burrows, Y.-K. Chen, R.W. Tkach, Compact and broadband coherent receiver front-end for complete demodulation of a 1.12-terabit/s multi-carrier PDM-QPSK signal, in *Proceedings of the European Conference on Optical Communications (ECOC)*, (2009), pp. 1–2
31. Z.B. Xu, R.Q. Hui, M. O'Sullivan, Dual-band OOFDM system based on tandem single-sideband modulation transmitter. Opt. Exp. **17**(16), 13479–13486 (2009)
32. Y. Tang, W. Shieh, Coherent optical OFDM transmission up to 1 Tb/s per channel. J. Lightwave Technol. **27**(16), 3511–3517 (2009)
33. W. Shieh, H. Bao, Y. Tang, Coherent optical OFDM: Theory and design. Opt. Exp. **16**, 841–859 (2008)
34. S.A. Mohammad, K. Maxim, L.J. Sander, W. Torsten, V.D.B. Dirk, D.W. Huug, et al., 11×224-Gb/s POLMUX-RZ-16QAM transmission over 670 Km of SSMF with 50-Ghz channel spacing. IEEE Photon. Technol. Lett. **22**(15), 1150–1152 (2010)

Chapter 4
Multi-band Fiber Transmission (MBT) Technology

Recently, the demand for higher capacity backbone networks, data centers, and optical access networks has increased dramatically owing to the popularity of 5G mobile communications and the transformation of online services driven by the impact of COVID-19 [1, 2]. Currently, the transmission capacity of single-mode optical fiber has reached 100 Tbits/s [3], which is close to the capacity limit of single-mode optical fiber transmission systems. The contradiction between the lack of growth in fiber optic transmission capacity and the bandwidth requirements of internet services will lead to a "capacity crunch". To expand the transmission capacity of coherent fiber optic communication systems, two available options are being investigated: space division multiplexing (SDM) and ultra-wideband (UWB) wavelength division multiplexing (WDM). For SDM, specialty fibers, such as multi-core fiber (MCF), few-mode fiber (FMF), and ring-core fiber (RCF) have been developed to enable multiplexing in several space and mode degrees of freedom [4–8]. This implies the deployment of new types of fibers and further research on transceivers and amplifiers for SDM. On the other hand, UWB WDM transmission is a technology that increases transmission capacity by extending the operating wavelength range of optical transmission systems to the entire low-loss spectral window of single-mode quartz fiber, i.e., O-, E-, S-, C-, L-, and U-bands [9]. By introducing transceivers and low-noise optical amplifiers that cover a wide wavelength range other than the Erbium-doped fiber amplifier (EDFA), a significant increase in fiber transmission capacity can be achieved without the need to redeploy the fiber transmission. This provides a practical solution for addressing the near-term "capacity crunch".

Many studies have demonstrated UWB transmission experiments and systems beyond the C + L band (see Table 4.1). The first WDM optical transmission experiment with 250 wavelength channels of 12.7 THz with SOA has been demonstrated; a net throughput of 115.9 Tb/s is transmitted over a 100 km SSMF with 49 GBaud probabilistic constellation shaping (PCS) of 64-ary quadrature amplitude modulation (QAM) [10]. In [11], a 45 GBaud polarization division multiplexing (PDM) 128QAM signal was used to achieve a record capacity of 150.3 Tb/s of a low

F. Tian, X. Xin, *High-Speed Large Capacity Optical Fiber Communications*,
Wireless Networks, https://doi.org/10.1007/978-3-031-94211-2_4

Table 4.1 Trends of the SMF transmission beyond the C + L band

Reference	[13]	[16]	[15]	[17]	[10]	[11]	[12]	[14]
Capacity (Tbit/s)	–	~35	46	99.35	115.9	150.3	178.08	256.4
Distance (km)	60	70	10072	257	100	40	40	54
Bandwidth (THz)	23.5	25.8	15.1	12.33	12.7	13.6	16.83	19.8
WDM band	O, E, S, C, L	E, S, C, L	S, C, L	S, C, L	S, C, L	S, C, L	S, C, L	S, C, L
Transmission fiber	LWPF	SSMF	SSMF	PSCF	SSMF	LWPF	Corning® SMF-28® ULL fiber	SSMF
Amplifier	SOA PDFA TDFA EDFA	DRA BDFA TDFA EDFA	TDFA EDFA	SOA DRA	SOA	EDFA TDFA	EDFA TDFA	DRA TDFA EDFA
Number of channels	625	148	552	247	250	272	660	793
WDM spacing (GHz)	37.5	100	25	50	50	100/50	25.5	25
Band rate (Gbaud)	32	30	24.5	49	49	45	25	24.5
Modulation format	QPSK 16QAM	16QAM	QPSK	16QAM PCS-64QAM	PCS-64QAM	128QAM	GS-64QAM GS-256QAM GS-1024QAM	256QAM

SSMF standard single mode fiber, *PSCF* pure silica core fiber, *SOA* semiconductor optical amplifier, *LWPF* low water peak fiber, *DRA* distributed Raman amplifier, *BDFA* bismuth-doped fiber amplifier, *TDFA* thulium-doped fiber amplifier, *PDFA* praseodymium-doped fiber amplifier, *EDFA* erbium-doped fiber amplifier, *PCS* probabilistically constellation shaped, *GS* geometrically shaped water peak fiber (LWPF) transmission over 40 km with 13.6 THz UWB, and an amplification scheme of EDFA and TDFA was adopted. In [12], hybrid distributed Raman/Doped fiber amplifiers were used to demonstrate the transmission of 660 × 25 GBaud geometric shaping (GS) QAM over 40 km, achieving a record throughput of 178.08-Tbits/s in a single-mode fiber network with 16.83 THz UWB. Moreover, a 5-band WDM transmission was demonstrated in [13]; the authors proposed an adaptive modulation format allocation of 100 G quadrature phase shift keying (QPSK) and 200 G 16QAM to exhaustively use the wavelength-dependent optical signal-to-noise ratio (OSNR). Recently, a 256.4 Tb/s transmission in the S-, C- and L-bands with 19.8 THz UWB were demonstrated over 54 km by using a 25 GHz-spaced PDM-256QAM signal [14]. Reference [15] achieved a 46 Tb/s transmission over 10,072 km with 15.1 THz UWB using doped fiber and distributed Raman amplification. In addition, a 25.8 THz UWB (E-, S-, C-, and L-bands) transmission using a multi-stage distributed Raman amplifier was achieved over 70 km by using 100 GHz-spaced 30 GBaud PDM-64QAM signals [16].

For UWB WDM transmission systems, the wavelength dependence (due to propagation loss) leads to capacity limitations. Additionally, as the system bandwidth increases, inter-channel stimulated Raman scattering (ISRS) plays an increasingly important role in limiting the capacity as a nonlinear impairment that redistributes optical power from short to long wavelengths during propagation [18]. Several models have been developed for nonlinear interference with the inclusion of ISRS [19–22], and various power optimization methods have been proposed to mitigate the partial depletion of short-wavelength channel power caused by ISRS [23–28]. In [23], the average fiber input power of the WDM signal in each S-, C-, and L-band is optimized by a sequential iterative search concerning the power tilt caused by ISRS. Following the sweep optimization of the average input power for single-band transmissions in the S and C bands, the L and S band power is optimized with the optimal power in the other bands, thus obtaining a high-capacity record. Meanwhile, the power pre-tilt and offset strategies in the S, C, and L-band scenarios are proposed [24, 25]. The local optimization global optimization (LOGO) algorithm is used to optimize the general signal-to-noise ratio (GSNR) by sweeping and combining the parameters of the launch power profile. The computational complexity is greatly reduced by optimizing the performance of a span transmission link to simplify the process. In [26], the simulated annealing algorithm is introduced to improve the LOGO strategy, allowing for a faster search for the optimal parameters of the transmit power; moreover, a higher transmission capacity can be obtained. In addition, a simple heuristic technique is used for individual band amplifiers in each S + C + L optical multiplexing section, and all amplifiers have static gain profiles that enable online compensation of span loss and adjustment of the launch power [27]. On the other hand, the scheme of embedding a Raman amplifier (RA) as an auxiliary amplifier after the fiber transmission has been proposed, and controlling the pumping wavelength and pumping power of the RA can compensate for short-wavelength signals with low signal power. Machine learning methods can be used to control the parameters of RAs. In [28], the proposed multi-band RA based on the machine learning framework uses up to 8 pumps to provide arbitrary gain over a 17.6 THz bandwidth in the S + C + L band. The power tilt caused by ISRS in UWB WDM systems will be effectively solved by this type of method.

In this chapter, we first introduce a transmission scheme using Fast-than-Nyquist rate in multi-band optical transmission system. A method of non-orthogonal transmission through duobinary filter is proposed and applied in multi-band optical transmission system. The simulation results show that FTN transmission can enhance the spectral efficiency of system, and we can restore signal via force zero equalization technology. Then, we investigate the performance of a UWB WDM transmission system that covers the widely used C + L band as well as the additional O-, E-, and S-bands. We establish the transmission system for UWB and discuss the effects of the channel, including Kerr nonlinearity and inter-channel interference from inter-channel stimulated Raman scattering (ISRS) between O-, E-, S-, C-, and L-bands. Moreover, we demonstrate an optimization scheme for compensating the spectral power tilt caused by SRS in the S + C + L band system, which utilizes the Raman amplifier and the partition particle swarm optimization (PPSO) algorithm. Finally, we demonstrate the ultra-wideband transmission of 480, 50 GHz spaced channels with PDM 256-QAM modulation over 20 km of SSMF. We achieved 115.2 Tbit/s transmission capacity by using artificial neural network to compensate fiber nonlinearity.

4.1 Fast-Than-Nyquist Rate (FTN) Technology

In recent years, the rapid growth of mobile and cloud services has rendered a strong demand for optical communication with high bit rate and low cost, and the requirements of optoelectronic devices for bandwidth and sampling rate have also increased correspondingly. Ultra-wideband (UWB) WDM transmission system has attracted much attention because the capacity in C band is limited. UWB scheme with C + L band has been introduced because the widely used EDFAs can be adopted in both bands [29]. Then bandwidth would be extended to the S band next, and finally five bands (the O, E, S, C, and L bands) would be utilized [30, 31].

Meanwhile, many methods have been proposed to solve the problem ISI in FTN scheme in the literature. When the sign probabilities of the sent sequences are equal, MLSE is a nonlinear algorithm with high complexity, which is generally implemented by Viterbi algorithm [32, 33]. However, the overall complexity of the algorithm increases exponentially as the modulation order and ISI truncation length are changing. Literature [31, 32] directly shortens the truncation length of ISI to reduce the complexity, but this will cause a certain loss of detection performance. Also, the ISI can be eliminated through the decoder and equalizer iteration at the receiver based on the Turbo iteration structure [33]. The generation process of ISI is derived from matrix operation, which combines simple inverse operation of matrix with decision feedback equalization (DFE) [34], which is suitable only for short detection sequence.

In this subsection, inter-symbol interference (ISI) is introduced artificially to compress symbol spacing because of the application of Faster-than-Nyquist (FTN) transmission technology in multi-band optical transmission system. The different modulation formats are used to achieve flexible signal transmission in different bands. Moreover, the method of FTN transmission through binary filter is introduced and applied to multi-band optical transmission system, which can improve the spectral efficiency effectively. At the receiver, the zero force (ZF) equalization technology, which is a linear low complexity detection algorithm, is used to recover the signal. In addition, different modulation formats are used to achieve flexible signal transmission in multi-band optical transmission systems.

4.1.1 System Model

A high spectral efficiency modulation scheme is designed for multi-band optical transmission system using the FTN technology as shown in Fig. 4.1. In addition, the optimal modulation format is geared to the channel band. The wavelength of the transmitted signal is set as 1310 nm and 1495 nm. The optical fiber transmission distance is 20 km, which shows that QPSK is allocated in O-band, 16QAM is distributed in the E and S bands, and 32QAM is used in the C and L bands. At the transmitter, the signals in different band channels are mapped into QPSK, 16QAM or 32QAM signals, and then the I-channel signals or Q-channel signals are respectively differentially encoded. After differential encoding, the signals are recombined into quantitative signals. The FTN signals of each band are obtained through the delay addition module.

The generation of a FTN signal can be achieved by filtering both signals using a low-pass filter with a bandwidth smaller than the Nyquist transmission, or by adding an optical bandpass filter smaller than the Nyquist transmission to the output of the optical domain modulator. FTN signals between different channels in each band are generated based on the duobinary delay addition filter.

Taking QPSK modulation as an example, Fig. 4.2 shows the implementation principle and the effect of the filter. In binary shaping, the symbol is added to the symbol delayed by one symbol period. It can be described by a double-tapped FIR filter with a rectangular Nyquist shaping. The impulse response is:

$$h_{duo}(t) = \frac{\mathrm{sinc}\left[\pi B\left(t - \frac{1}{2B}\right)\right] + \mathrm{sinc}\left[\pi B\left(t + \frac{1}{2B}\right)\right]}{2} \tag{3.50}$$

where t is time and B is signal baud rate. The impulse response consists of the delayed impulse response of two rectangular Nyquist shapes. The amplitude spectrum of the signal after binary shaping is expressed as:

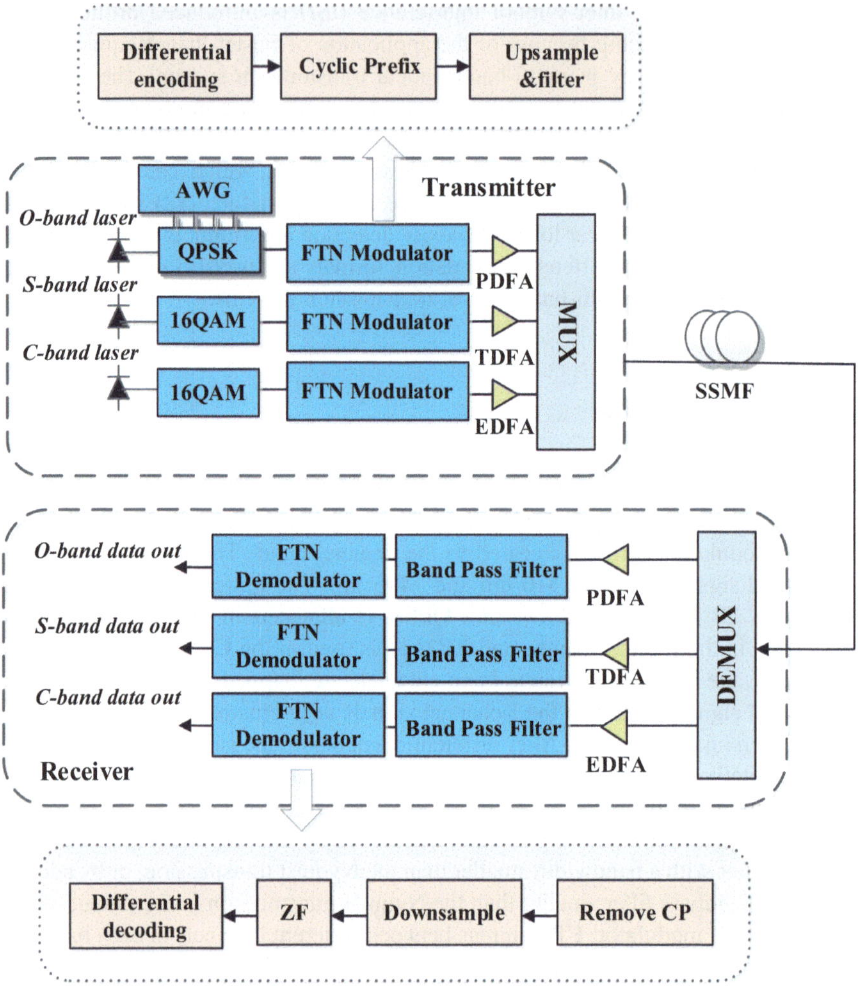

Fig. 4.1 The FTN multi-band optical transmission system

$$H_{duo}(f) = \begin{cases} 0 & \text{for } |f| \geq B/2 \\ \cos\left(\frac{\pi f}{B}\right) & \text{for } |f| < B/2 \end{cases} \tag{3.51}$$

After multi-binary shaping, the signal spectrum is in the shape of cosine, and the bandwidth is 1/2 of the original signal. The above multi-binary shaping should be maintained for the total transfer function consisting of transmitter, receiver, and transmitting channel. In addition, the matched filter configuration achieves the best BER in the channel of AWGN.

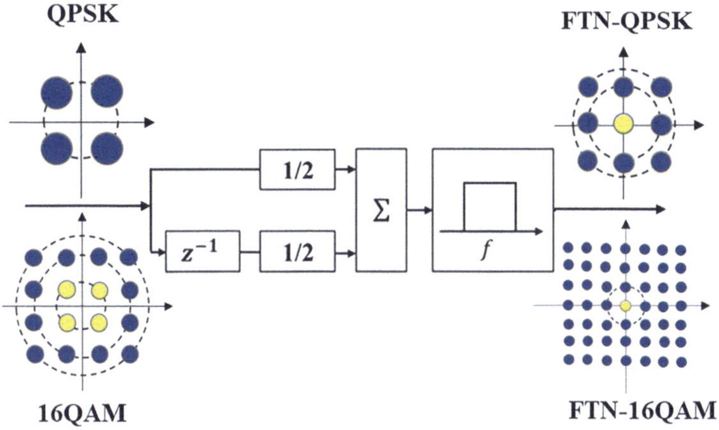

Fig. 4.2 Schematic configuration of duobinary shaping

As a result of FTN filtering, the constellation diagram of QPSK signal will change from 4 points to 9 constellation points, which is similar to 9QAM signal, and 16 points of 16QAM signal will change to 49 constellation points, which is similar to 49QAM signal. In this section, the FTN signal generation is based on the addition of two binary delay filter element before and after the operation process, the signal, the receiving end to restore the original binary signal, requiring the transmitter to differential signal encoding. Then signal transmit through the delay additive module. A FTN signal is generated.

At the receiver, the scheme of the multi-band optical transmission system uses FTN demodulation and equalization techniques to recover the original data. The transmission signal passes through the wavelength division multiplexer, and the non-orthogonalized overlapping optical signals of each band are multiplexed in frequency domain to generate frequency domain multiplexed signals. After signals transmitting through the optical fiber channel, digital signal processing is carried out in each wave band by wave decomposition multiplexer, likewise, the corresponding differential decoding is required at the receiving end. The differential decoding process can facilitate the reception of the FTN signal after the tough decision. After removing the cyclic prefix, the down sampling and zero forced equalization are carried out. Finally, the signal after differential decoding at the receiver is demodulated by QAM to realize the data output.

4.1.2 Results and Discussion

The performance of the FTN transmission technology applied to the O-band signal transmission is shown in Fig. 4.3. It is assumed that the RRC rolling factor is α, the FTN time compression factor is τ and the total code rate is R, so the system spectral

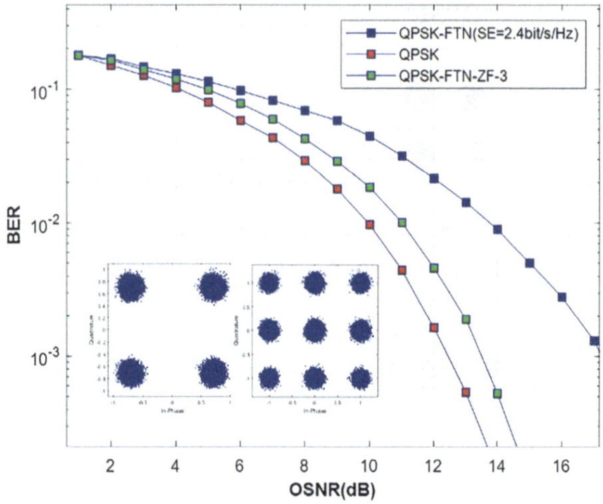

Fig. 4.3 BER performance in FTN-QPSK in O band (L = 20 km)

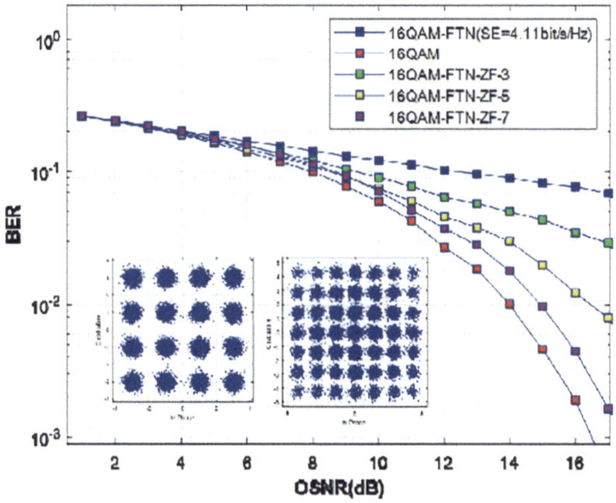

Fig. 4.4 BER performance in FTN-16QAM in S band (L = 20 km)

efficiency (SE) is obtained $\eta = mR/\tau(1 + \alpha)$, where m is the number of quantization bits of a symbol. In the QPSK-FTN transmission system, m is 2, and RRC rolling factor is 0.1. It is calculated that the spectral efficiency of FTN-QPSK is 2.4 bit/s/Hz. The spectral efficiency of FTN-16QAM is 4.11 bit/s/Hz.

When the OSNR is set to 14 dB, the BER of QPSK is 1.4×10^{-4}, and the BER of that is 8.97×10^{-3} with the addition of the FTN filter. While BER performance can be improved to 5.3×10^{-4} by applying zero force equalization at the receiver, which can eliminate part of inter-symbol interference for FTN. Figure 4.4 shows the BER performance in the C-band using FTN-16QAM. After FTN modulation, 16 points of 16QAM become 49 constellation points. When the OSNR is set to 16dB, the BER of 16QAM signal is 1.9×10^{-3}. The BER of that is 7.76×10^{-2} with the addition of the FTN filter. While BER performance can be improved by applying zero force equalization at the receiver, which can eliminate part of inter-symbol interference for FTN better by increasing the number of transversal filters. The zero-force equalization of the seventh-order transverse filter greatly improves the BER performance of FTN-16QAM.

In this subsection, we demonstrate the FTN scheme used in multi-band optical transmission system. The adaptive modulation format allocation is proposed to achieve the maximum capacity transmission. The spectral efficiency of QPSK modulation in O-band is increased to 2.4bit/s/Hz, and the spectral efficiency in S-band is increased to 4.11bit/s/Hz. At the receiver, signals are recovered by ZF equalization algorithm to reduce the ISI effectively. The results show that the combination of FTN and multi-band optical transmission system has sound application prospect in future large capacity optical communication.

4.2 Performance Analysis and Power Tilt Mitigation of Ultra-Wideband WDM Transmission Systems

In this subsection, a UWB WDM system is built for investigating the impacts of ISRS on the transmission performance between O-, E-, S-, C-, and L-bands with the consideration of Kerr nonlinearity. To further mitigate channel impacts and improve spectral efficiency, an optimization scheme to compensate for the power tilt is demonstrated using a RA, and a partition particle swarm optimization (PPSO) algorithm is introduced to secure the optimal pump power and pump wavelength for the RA when the output signal power spectrum is the flattest. The results show that the power tilt is reduced from 18 to 2.93 dB. The iteration speed is increased by 10% compared to the normal PSO algorithm.

4.2.1 Analysis of the UWB WDM Transmission System

4.2.1.1 System Model and Simulation Setup

To investigate the effect of nonlinearity on the signal performance of UWB WDM systems, the nonlinear Schrödinger equation for optical pulse transmission in a single-mode fiber is first given as [35]:

$$\frac{\partial E_i}{\partial z} = \frac{-\alpha}{2}E_i - \frac{j\beta_2}{2}\frac{\partial^2 E_i}{\partial T^2} + \frac{\beta_3}{6}\frac{\partial^3 E_i}{\partial T^3} + j\gamma\left[\left|E_i\right|^2 + 2\sum_{k\neq i}^{N}\left|E_k\right|^2\right]E_i \tag{4.1}$$

where E_i is the slow-varying envelope of the i-th channel optical field, z is the transmission distance, N is the number of channels, α is the attenuation coefficient, β_2 is the chromatic dispersion (CD) parameter, β_3 is the dispersion slope, and γ is a

Table 4.2 Parameters of different bands [9]

Band	Wavelength (nm)	Frequency (THz)	Bandwidth (THz)	Nonlinear coefficient (1/W/km)	Attenuation coefficient (dB/km)	CD coefficient (ps/nm/km)
O	1260–1360	220.59–238.10	17.25	1.6	0.38–0.28	−5–4
E	1360–1460	205.48–220.59	14.81	1.5	0.28–0.22	4–12
S	1460–1530	196.08–205.48	9.13	1.4	0.22–0.18	12–16
C	1530–1565	191.69–196.08	4.13	1.3	0.18	16–18
L	1565–1625	184.62–191.69	6.96	1.28	0.18	18–22

nonlinear coefficient. The first term on the right side of (Eq. 4.1) represents the loss of the optical field. The second and third terms represent the CD and dispersion slope of the fiber, while the third term is the phase change caused by nonlinear impairments, including the self-phase modulation (SPM) and cross-phase modulation (XPM). In UWB WDM systems, wavelength-dependent channel characteristics have to be coped with dispersion coefficient ranging from −5 ps/nm/km to 22 ps/nm/km; fiber loss α is 0.18 dB/km to 0.38 dB/km; and the nonlinear coefficient γ is 1.28 1/W/km to 1.6 1/W/km. Table 4.2 shows the attenuation and CD coefficients for the five bands (O→L-band) of the ITU-T G.652.D fiber considered constants.

Subsequently, a UWB WDM transmission system covering O-, E-, S-, C-, and L-bands was built in VPIphotonics based on the wavelength-dependent channel characteristics, and the nonlinear Schrödinger equation was solved by the split-step Fourier method (SSFM), commonly used for numerically solving the nonlinear Schrödinger equation and simulating WDM signal transmissions. The simulation setup is shown in Fig. 4.5. At the transmitter, 100 continuous wave (CW) light sources were placed with a spacing grid of 100 GHz for WDM transmission in O and E bands. In the S-, C-, and L-bands, 90, 40, and 70 CW sources were placed respectively. It should be noted that the simulation time of SSMF can be substantial even with fast computing hardware, especially for UWB WDM transmission beyond the C + L band. Therefore, to speed up the system computation process, the signal was modulated to the carriers of five channels in each band, and the carriers of the other channels were unloaded to reduce the computational complexity. The CW light of five channels in five bands were emitted to the IQ modulator (IQM) of each band with the linewidths of 100 kHz. The arbitrary waveform generator (AWG) was used to drive the IQM for the 32 GBaud polarization division multiplexing (PDM) signal modulation. The signals were root-raised cosine-shaped with a roll-off of 0.05. The light of the modulated channels and unmodulated channels were coupled together in each band. A multiplexer (MUX) was then used to combine the signals of the five

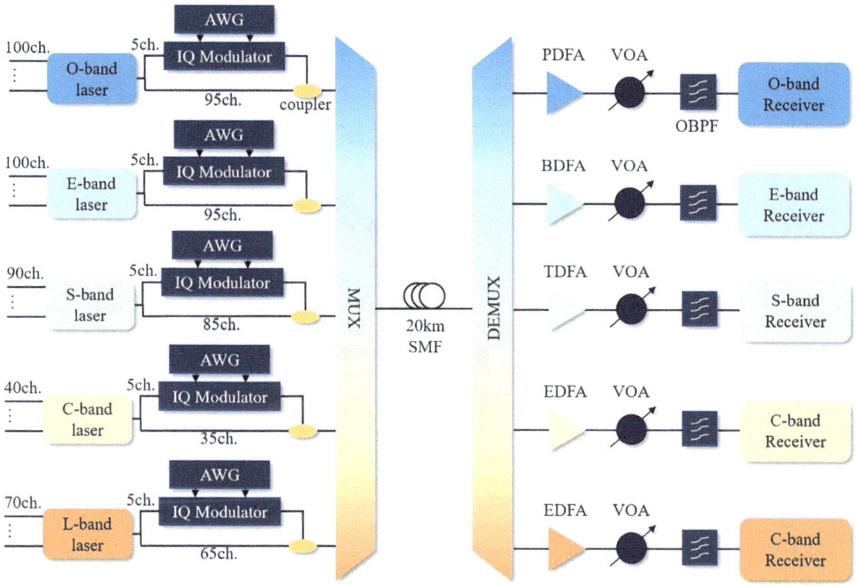

Fig. 4.5 System schematic for the 5-band WDM transmission

bands. After 20-km SMF transmission, the modulated WDM signals were separated by a de-multiplexer (DEMUX) for each band. Subsequently, the signal lights were amplified by the praseodymium-doped fiber amplifier (PDFA) in the O-band, BDFA in the E-band, thulium-doped fiber amplifiers (TDFAs) in the S-band, and EDFA in the C + L band respectively. A variable optical attenuator (VOA) was used for power control, and an optical band-pass filter (OBPF) was to filter light beyond the desired wavelength. At the receiver, band-ideal coherent receivers were used to detect the signal, which was then fed into digital signal processing (DSP) to compensate for the impairment. The receiver DSP consisted of the Gram–Schmidt orthogonalization procedure (GSOP), digital back-propagation (DBP), the constant modulus algorithm (CMA), and carrier recovery.

4.2.1.2 Five-Band Individual Channel Performance Analysis and Modulation Format Assignment

First, we investigated the performances of individual channels in each band and assigned the appropriate modulation format in each band. The dual polarization optical signal-to-noise ratio (OSNR) can be expressed as [36]

$$\text{OSNR} = \frac{R_s}{B_{ref}}\text{SNR} \tag{4.2}$$

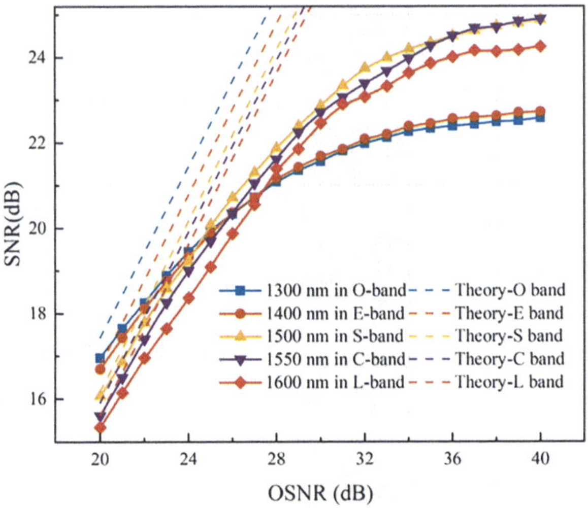

Fig. 4.6 SNR versus OSNR for five wavelengths in different bands

where R_s is the symbol rate and B_{ref} is the reference bandwidth. The reference bandwidths in this section are 17.7 GHz, 15.3 GHz, 13.3 GHz, 12.5 GHz, and 11.7 GHz (O → L) respectively. In Fig. 4.6, the SNR (in the five channels as a function of OSNR) is shown at 1300 nm, 1400 nm, 1500 nm, 1550 nm, and 1600 nm. With an OSNR of more than 36 dB, there is a ceiling of 22.5 dB and 22.7 dB for the O/E-band respectively. For the S-, C-, and L-bands, the SNR peaks at approximately OSNR = 38 with 24.9 dB, 24.9 dB, and 24.2 dB respectively. The SNR gap versus the theoretical, as well as the SNR ceiling, are caused by the amplified spontaneous emission (ASE) noise from the band-dependent optical amplifier and the transceiver noise. Based on the resulting SNR, we chose the PDM-16QAM or PDM-32QAM modulation schemes to reduce the excess margins for different bands. Assuming a BER threshold of 3.8×10^{-3} for FEC, the SNR at the FEC threshold should theoretically be about 16.5 dB at 16QAM and about 19.2 dB at 32QAM. In consideration of the SNR penalty in the WDM system, we allocated 16QAM to the O and E bands and 32QAM to the S-, C-, and L-bands to ensure an adequate SNR window. Fig. 4.7 illustrates the OSNR of the individual 5-wavelength channels as a function of launch power. The OSNR of the channel is maximized when the launch power reaches 2 dBm, and then the OSNR of the system decreases due to the enhanced nonlinear interference. Moreover, the OSNR of the O/E-band is from 29.8 dB to 37.8 dB, and the OSNR of the S-, C-, and L-bands 34.5 dB to 43.9 dB, which is in accordance with the requirements for the assigned modulation format.

Figure 4.8 shows the BER for each band as a function of the OSNR of each channel. With the lower modulation order, the BER of 16QAM signals in the O/E-band is generally lower than that of 32QAM signals in the S-, C-, and L-bands. When the OSNR is increased to 30 dB, all five-band channels are available at the FEC threshold. Figure 4.9 shows the BER for each band as a function of launch power. The BER of the five channels reaches their respective minima from

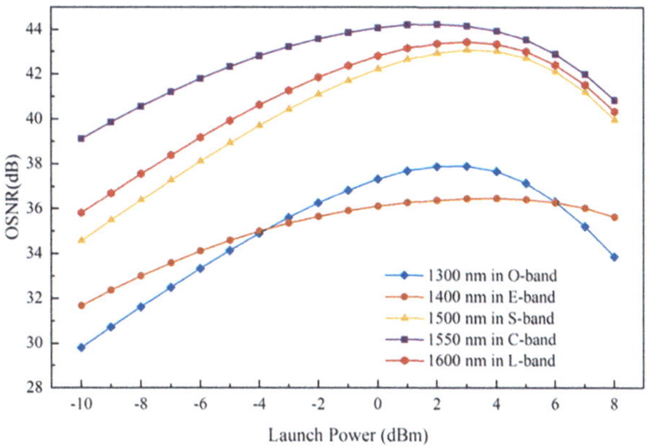

Fig. 4.7 OSNR versus launch power for five wavelengths in different bands.

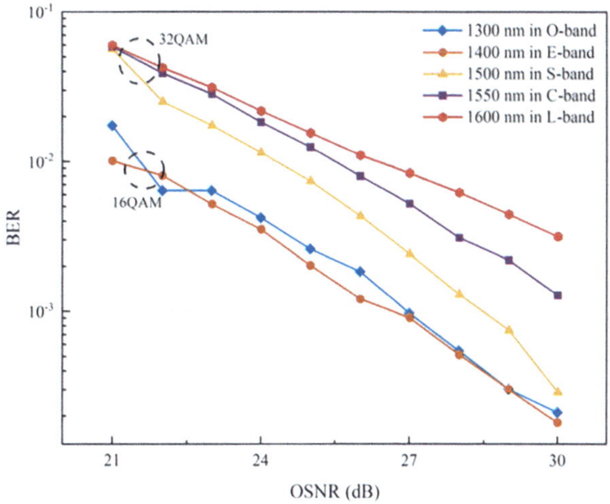

Fig. 4.8 BER of individual channels in five bands as a function of OSNR.

6 to 8 dBm, while the OSNR shown in Fig. 4.7 achieves the respective maxima at 2 dBm. This is because, at the launch power < 6 dBm, the contribution of the launch power boost to SNR is greater than that the performance degradation from nonlinearity. When the launch power > 8 dBm, the nonlinear degradation dominates, and the system performance degrades. Additionally, it can be seen that the C-band performance is optimal, and the launch power increase is the most significant for the C-band performance improvement, which is consistent with the results in Fig. 4.7.

4.2.1.3 Results of the Five-Band WDM Transmission

To evaluate the performance penalty of the five-Band WDM system transmission compared with the single-carrier transmission, the BER and generalized mutual

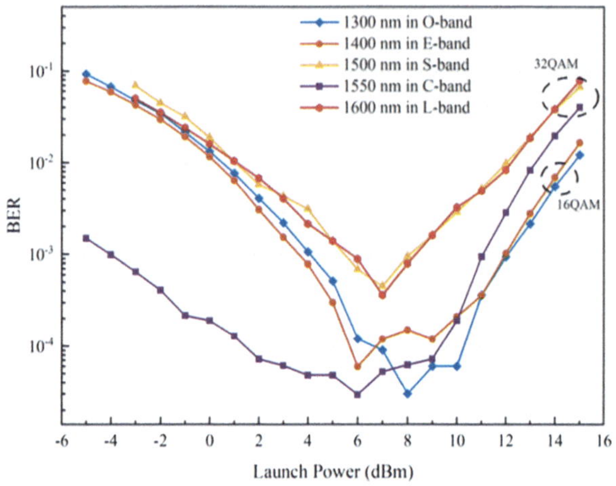

Fig. 4.9 BER of individual channels in five bands as a function of launch power.

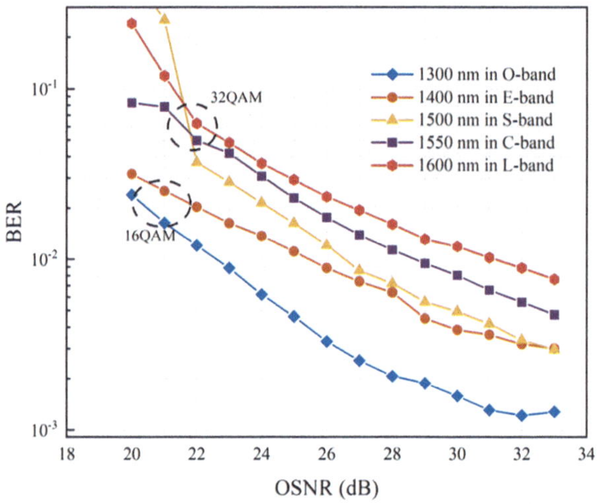

Fig. 4.10 The BER of the test channel as a function of OSNR in five-band WDM transmission

information (GMI) are first calculated for the five test channels. The presence or absence of crosstalk essentially depends on the WDM frequency spacing, and inter-channel crosstalk becomes important as the channel spacing approaches the symbol rate. In our setup, the WDM frequency spacing is set to 100 GHz, and the impact of crosstalk is minimal or even negligible [36]. This implies that the WDM penalty introduced by the five-band system is primarily from inter-channel nonlinearity. Fig. 4.10 illustrates the BER of the test channel as a function of OSNR in five-band WDM transmission. The OSNR penalty of the E-band test channel is significantly larger in the five bands compared to the results in Fig. 4.6, which suggests that the signal will be dominated by nonlinear interference in the E-band. The GMI versus OSNR for the five-band test channels is given in Fig. 4.11. The value of GMI gives the maximum number of bits of information transmitted with the error vanishing

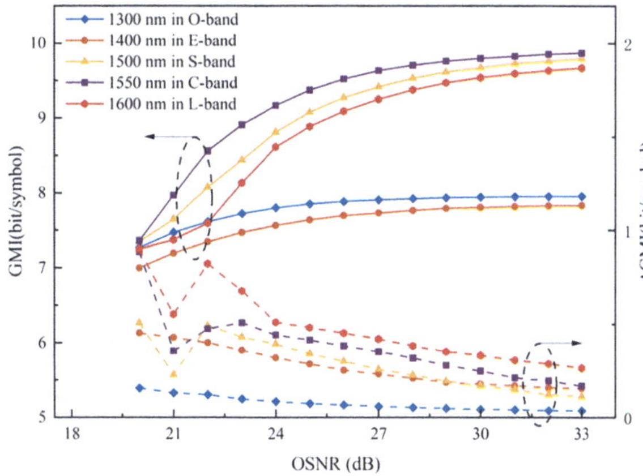

Fig. 4.11 The GMI and ΔGMI of the test channel as a function of OSNR in five-band WDM transmission

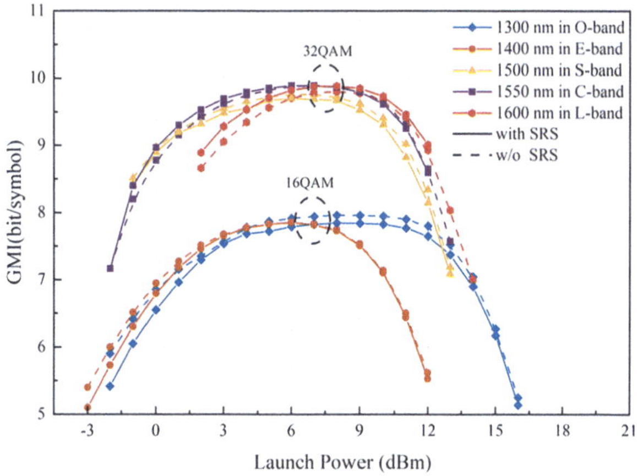

Fig. 4.12 The GMI of the test channel with and without ISRS as a function of launch power in five-band WDM transmission

probability and can be used to estimate the system throughput. The dashed line represents the GMI gap for WDM transmission, defined as $\Delta GMI = GMI_{Single} - GMI_{WDM}$. The test channel of the O-band features the lowest ΔGMI, even with the highest γ in the five bands. It is predicted that there is a correlation between the high noise tolerance of 16QAM assigned by the O-band and the higher noise factor of the optical amplifier, together with the transmission distance of 20 km, where the accumulated nonlinear interference is not dominant. For the S-, C-, the L-bands, and the ΔGMIs are higher due to the lower noise tolerance of 32QAM, with the penalty being the highest for the L-band. Figure 4.12 shows the GMI as a function of the launch power for five-band WDM transmission. For the O/E-band, there is a narrower high GMI window for the E band due to nonlinear interference. Likewise, the test channel at 1600 nm in the S, C, and L-band features a smaller window of high GMI due to the higher penalty. The launch power of each band reaching the maximum GMI has different offsets. Considering the power consumption, we choose the launch power when the GMI just reaches the ceiling with the optimal launch power for the five-band WDM system, and the optimal launch powers for the O to L-band are 7, 5, 5, 6, and 7 dBm, respectively.

Furthermore, the effect of ISRS becomes non-negligible as the bandwidth of the WDM signal increases, especially in WDM systems beyond the C + L band. ISRS effectively causes energy transfer from shorter wavelengths to longer wavelengths, peaking at 100 nm. The coupling between the WDM ith channel and all other channels, considering ISRS and fiber attenuation, is described as [37]

$$\frac{\partial P_i}{\partial z} = -\alpha(f_i)P_i - \sum_{j=1}^{i-1} \frac{f_i}{f_j} g_R(|f_i - f_j|) + \sum_{j=i+1}^{N} g_R(|f_i - f_j|) \tag{4.3}$$

where P_i represents the power of the ith channel, f_i is the frequency of the ith channel, $\alpha(f_i)$ is the fiber attenuation profile in the absence of ISRS, and $k = 1$ indicates the channel with the lowest frequency. The first term on the right-hand side of (4.3) describes the fiber attenuation of the i-th channel, the second term describes the power transfer from the i-th channel to the lower frequency channel, and the third term illustrates the power transfer from the higher frequency channel to the i-th channel. It is assumed that the Raman gain coefficient g_{R} varies negligibly over the bandwidth of each channel. The nonlinear Schrödinger equation is still solved numerically by SSFM, considering the effect of ISRS. The wavelengths we chose for the WDM test channels were 1300, 1400, 1500, 1550, and 1600 nm, which are the central wavelengths of each band. They are mostly subject to interference from in-band nonlinearity and feature a wavelength difference of around 100 nm, providing the strongest ISRS. It has been shown that the crosstalk penalty originating from the amplitude fluctuations of the interaction channel is negligible [38, 39]. Therefore, we focus exclusively on the ISRS-induced optical power tilt. Figure 4.13 shows the optical power transfer of each channel. Due to the aggregated power transfer effect of the five bands, the short-wavelength channels in the O-, E-, and S-bands provide power gain for the C and L-band signals as pump channels while being partially depleted. While the quality of the long-wavelength channels is guaranteed, the short-wavelength channels may be underpowered, resulting in a decrease in system throughput.

The relationship between GMI and launch power in the presence of ISRS is shown in Fig. 4.12. The O/E/S band experiencing GMI penalties for each test channel is compared to the absence of ISRS, while the C- and L-band acquire GMI gain, which is consistent with the results in Fig. 4.13. Simultaneously, there is no effect of SRS on the launch power value of each test channel reaching the GMI ceiling. The relationship between SNR and GMI versus wavelength for five modulated channels per band in the five-band WDM transmission at optimal launch power is shown in Fig. 4.14. The curves of OSNR and GMI either with or without ISRS intersect in the C-band, which is consistent with the results in Fig. 4.13 that the C- and L-bands acquired performance gain in the O/E/S-band via ISRS. At the optimal launch power, the GMI of the test channel in the O/E/S-band decreases by 0.2 bits/symbol, 0.08 bits/symbol and 0.09 bits/symbol respectively, while the GMI of the

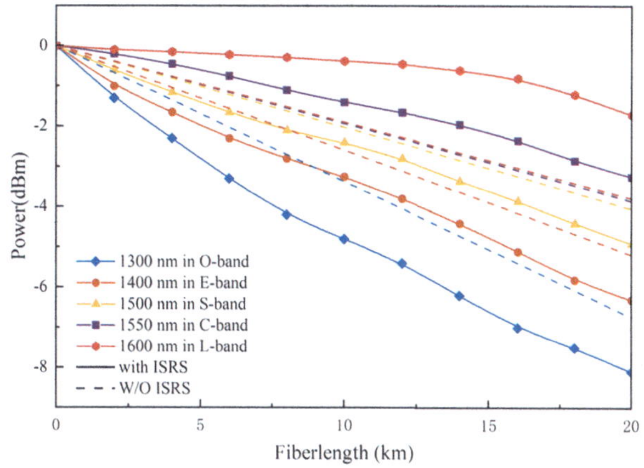

Fig. 4.13 Power transfer in the 5-band transmission

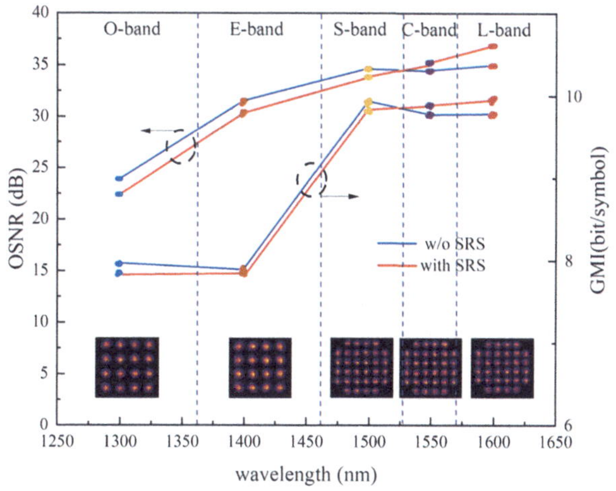

Fig. 4.14 SNR and GMI vs. wavelength in 5-band WDM at optimal launch power. constellation of each band shown below

test channel in the C- and L-bands increases by 0.09 bits/symbol and 0.1 bits/symbol respectively. Since we only have five test channels for modulated signals in each band, the throughput is roughly evaluated by considering each band. The throughput can be expressed as

$$T = \sum (N_{band} * - GMI_{band} * R_s) \tag{4.4}$$

where T is the throughput, N_{band} is the number of channels per band, and GMI_{band} is the average GMI of the test channels for each band. Without ISRS, the GMI_{band} of the O → L band test channels in Fig. 4.14 are 7.94 bits/symbol, 7.87 bits/symbol, 9.89 bits/symbol, 9.79 bits/symbol, 9.69 bits/symbol, and the system throughput is 113.312 Tbits/s. Considering ISRS, the total system throughput is 112.496 Tbits/s, resulting in a throughput reduction of 816 Gbit/s.

4.2.1.4 Conclusions and Discussion

We demonstrate a five-band UWB WDM system with VPIphotonics by analyzing the single-wavelength performance per band and assigning an appropriate modulation format based on the SNR margin of the band. In addition, the five-band WDM transmission is demonstrated, and the launch power of each band is optimized on a band basis in this configuration. ISRS in the UWB WDM system is analyzed to evaluate the impact on the system, and it results in a system throughput degradation of 816 Gbit/s, ultimately leading to a total throughput of 112.496 Tbits/s. The transmission distance is limited to 20 km due to the cumulative nonlinearity and the computational complexity of the model. Therefore, in short-distance scenarios such as data center interconnects, where there is a high demand for dense network traffic and large-capacity UWB WDM transmission. It is expected that UWB WDM transmission will be applied first in this area.

In the simulation research, the total system bandwidth of over 50 THz, renders a sharp increase in computational complexity and a system simulation time of up to 20h even with a high-performance graphics processing unit (GPU). More efficient and fast methods can be introduced to evaluate the system performance for UWB WDM transmission, such as Gaussian noise (GN) models, which will be our focus for future work.

In addition, it is necessary to adequately cope with the impact of ISRS in the design of UWB WDM systems to reduce the throughput penalty. Several studies have been conducted to address the power tilt due to ISRS by optimizing the launch power profile or adding complementary fiber gratings. In Sect. 4.3, we focus on the power tilt compensation in S-, C-, and L-bands by introducing a RA as an auxiliary amplifier after fiber transmission to achieve simple and fast compensation of the power tilt by controlling the pumping wavelength and pumping power.

4.2.2 Compensation Scheme for Spectral Power Tilt

4.2.2.1 Compensation Scheme for ISRS

The gain spectrum of the RA is entirely determined by its pump wavelengths and pump power. Therefore, a channel of any wavelength can be amplified by a RA with a proper pump source. Our compensation scheme is demonstrated in the S-, C-, and L-bands of the UWB WDM system is depicted in Sect. 4.2.1. The optimization setup for the S-, C- and L-bands in the UWB WDM system are shown in Fig. 4.15. A RA is added after the fiber transmission, and the pump source of the RA is connected to a control unit, which introduces a PPSO algorithm to control the pump wavelength and pump power of the RA according to the flatness of the output power spectrum. The pseudo-code of the PPSO algorithm is shown in Algorithm 4.1.

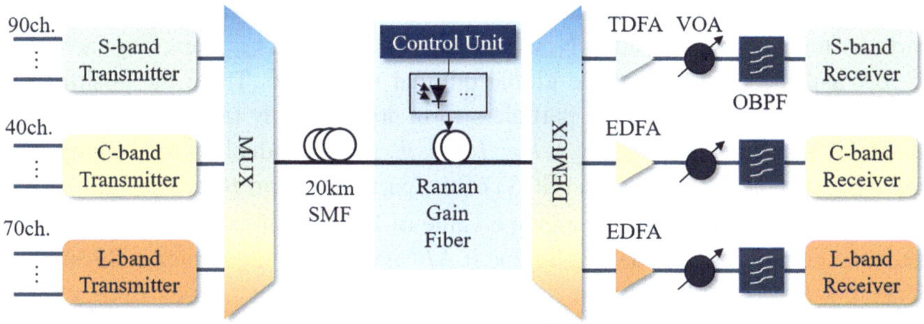

Fig. 4.15 The optimization setup for the S-, C-, and L-bands in the UWB WDM system

Algorithm 4.1 Partitioned Particle Swarm Optimization

Initialize v_k, x_k, $k = 1,\ldots,m$
Compute $f(x_{k,0})$, $B_k^p \leftarrow x_k k = 1,\ldots,m$
for $i = 1$ to N **do**
  Compute $f(x_{k,i})$, $k = 1,\ldots,m$
  Compute $\overline{f(x_i)} \leftarrow \text{mean}\{f(x_{k,i})\}$
  Compute B_t^g
  Compute $\Delta B^g \leftarrow B^g - B_t^g$
  if $\Delta B^g < \delta$ **then**
    $c \leftarrow c + 1$
  end if
  if $c > c_{\max}$ **then**
    if $f(x_{k,i}) < f(x_{k,i-1})$, $k = 1,\ldots,m$ **then**
      $x_{k,i} \leftarrow x_{k,i-1}$ Gaussian(1,1)
    else
      $x_{k,i} \leftarrow x_{k,i-1}$
    end if
  else
    if $f(x_{k,i}) > \overline{f(x_i)}, \quad k = 1, \ldots, m$ **then**
      $n_t \leftarrow (t_{\max} - t)/t$
      $x_{k,i} \leftarrow x_{k,i-1}[1 + n_t C(0,1)]$
    else
      $v_{k,i} \leftarrow c_1 v_{k,i-1} + c_2 \text{rand}\,(B^p{}_{k,i-1} - v_{k,i-1})$
        $+ c_3 \text{rand}\,(B^g{}_{k,i-1} g - v_{k,i-1})$
      $x_{k,i} \leftarrow x_{k,i-1} + v_i$
    end if
  end if
end for

Firstly, each parameter is bounded, including the particle swarm D, the number of particles m, maximum iteration N, velocity bounds v_{min} and v_{max}, learning factors c1, c2, c3, and the threshold of the global optimal solution δ. Then, the position and velocity of each particle in the particle swarm are randomly initialized. The fitness value (FV) of each particle $f(x_{k,\ i})$, $k = 1, \ldots, m$, the individual optimal solution B_k^p, and the global optimal solution (GOS) of the particle swarm B_k^g are calculated. The variation of GOS ΔB^g and the average value of FV $\overline{f(x_t)}$ also need to be obtained. The count c will be incremented by one if ΔB^g is less than the variation threshold δ. When $c > c_{max}$, the optimization is considered to be caught in the local optimum, and then strategy 1 is introduced to move out of the local optimum. On the contrary, strategy 2 is used to update.

Strategy 1. The positions of the particles are updated by adding a Gaussian disturbance to avoid falling into the local optimum. The updated formula is as follows:

$$x_{k,i} = \begin{cases} x_{k,i-1}\text{Gaussian}(1,1) & (f(x_{k,i}) < f(x_{k,i-1})) \\ x_{k,i-1} & (f(x_{k,i}) \geq f(x_{k,i-1})) \end{cases} \tag{4.5}$$

where Gaussian (1, 1) denotes a Gaussian function with a mean of 1 and variance of 1. If the fitness of the updated parameter is less than that of the original, the parameters of the swarm will be updated according to (4.3). Otherwise, the original parameter array remains unchanged.

Strategy 2. The particles are grouped according to their fitness values. For $f(x_{k,i}) < \overline{f(x_l)}$, the particle is considered a preferred particle, while for $f(x_{k,i}) \geq \overline{f(x_l)}$, the particle is considered a sub-optimal particle. For the preferred particles, the Cauchy formula is introduced to update the parameters and more particles are generated in the preferred region to enhance the global detection of the particles. The formula is as follows:

$$\begin{aligned} X_{i+1} &= X_i \cdot [1 + n_t \cdot C(0,1)] \\ n_t &= \frac{t_{max} - t}{t_{max}} \end{aligned} \tag{4.6}$$

where $C(0, 1)$ is the random number generated by the Cauchy distribution function, $t_{\max}$ is the maximum number of iterations, and t is the current number of iterations. For the sub-optimal particles, the particles should be accelerated to the evolution of the preferred particles, and the parameters are updated by the following formula:

$$\begin{aligned} v_{k,i} = {} & c_1 v_{k,i-1} + c_2 \text{rand} \cdot \left(B_{k,i-1}^p - v_{k,i-1}\right) \\ & + c_3 \text{rand} \cdot \left(B_{k,i-1}^g - v_{k,i-1}\right) \end{aligned} \tag{4.7}$$

$$x_{k,i} = x_{k,i-1} + v_i \tag{4.8}$$

where it denotes a random value from 0 to 1.

Note that the particle update speed and position should be limited. If the velocity or the position of the particle exceeds the set range, the boundary value is used as the updated value. When the maximum number of iterations is reached, the optimization is terminated after all particles are being updated, and the optimal positions of the particles and the global optimal solution are produced as outputs.

For this algorithm, the population is dynamically divided into two subgroups, and different evolutionary mechanisms are designed for the particles in each subgroup, which can enhance the diversity of the population and accelerate the speed of obtaining the global optimal solution. Meanwhile, the algorithm tackle the problem of being trapped in the local optimum by adding a Gaussian interference.

4.2.2.2 Simulation and Results

In the simulation, the optimal launch power values for the S, C, and L-band WDM transmissions are 4, 5, and 6 dBm, respectively, by using the method in Sect. 4.2. The power optimization scheme based on RA with PSO (Raman-PSO) or RA with PPSO (Raman-PPSO) is investigated for both the 3 dBm launch power scenario and the scenario with optimized launch powers. The pump wavelength and pump power are employed as optimization parameters, and the mean squared deviation of the output signal spectrum is considered the particle fitness. The exact form of fitness can be expressed as

$$\sigma = \sqrt{\frac{\sum_{i=1}^{n} [X_I - E(X)]^2}{n}} \tag{4.9}$$

where n is the initial number of particles, X_I is the output signal power, and E(X) is the average value of the output signal spectrum. The parameters of the algorithm and the optimization parameters of the two power optimization schemes are given in Tables 4.3 and 4.4 respectively.

In Fig. 4.16, the yellow carve indicates the tilt in the output power spectrum due to the transfer of the signal power from the S-band to the L-band induced by the ISRS. We define the power tilt $\Delta P = P_{max} - P_{min}$. As launch power = 3 dBm, the unoptimized ΔP is 9.55 dB and the ΔPs are 1.93 dB and 1.44 dB after optimization by both Raman-PSO and Raman-PPSO schemes respectively. The latter scheme can effectively compensate for ISRS and it provides a 0.49 dB improvement over the

Table 4.3 The parameters of the PPSO algorithm

Parameter	Value	Parameter	Value
D	30	x_{min}	1579 nm
m	20	c_1	0.3
v_{max}	10	c_2	1.5
v_{min}	−10	c_3	1.5
x_{max}	1250 nm	δ	0.1

Table 4.4 The parameters of the PPSO algorithm

	Raman-PSO		Raman-PPSO	
	Pump wavelength (nm)	Pump power (dBm)	Pump wavelength (nm)	Pump power (dBm)
Figure 4.16a	1476.7	23	1498.8	22.6
	1419.6	20	1367.8	24.2
	1441	21.2	1494.9	19.2
Figure 4.16b	1466.2	22	1375.2	23.2
	1409.1	21.5	1498.6	22.4
	1388.2	20.1	1455	20.1

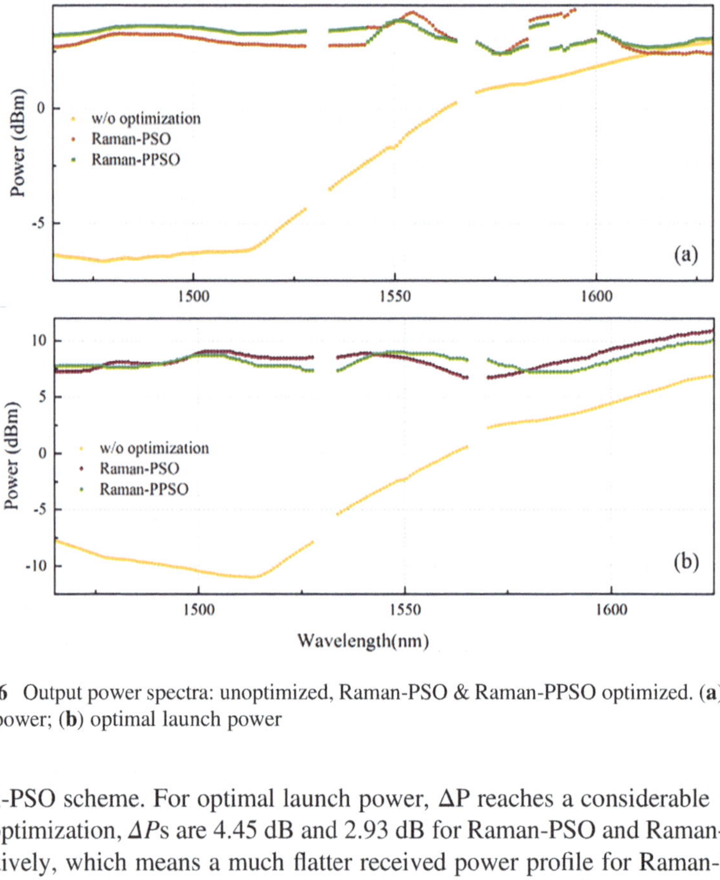

Fig. 4.16 Output power spectra: unoptimized, Raman-PSO & Raman-PPSO optimized. (**a**) 3 dBm launch power; (**b**) optimal launch power

Raman-PSO scheme. For optimal launch power, ΔP reaches a considerable 18 dB. After optimization, ΔPs are 4.45 dB and 2.93 dB for Raman-PSO and Raman-PPSO respectively, which means a much flatter received power profile for Raman-PPSO. Figure 4.17 shows the fitness value as a function of the number of iterations, the proposed Raman-PPSO scheme iterates 10% faster than Raman-PPSO, which indicates more efficient power tilt mitigation.

To evaluate the system transmission performance, three wavelength channels from the S-, C-, and L-bands were received: 1496, 1550, and 1600 nm. Figure 4.18 shows the system performance without optimization, with Raman-PSO and Raman-

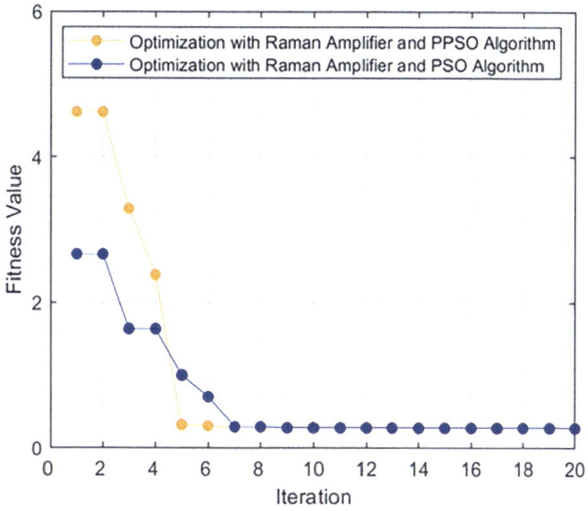

Fig. 4.17 Fitness value versus iteration for optimization schemes

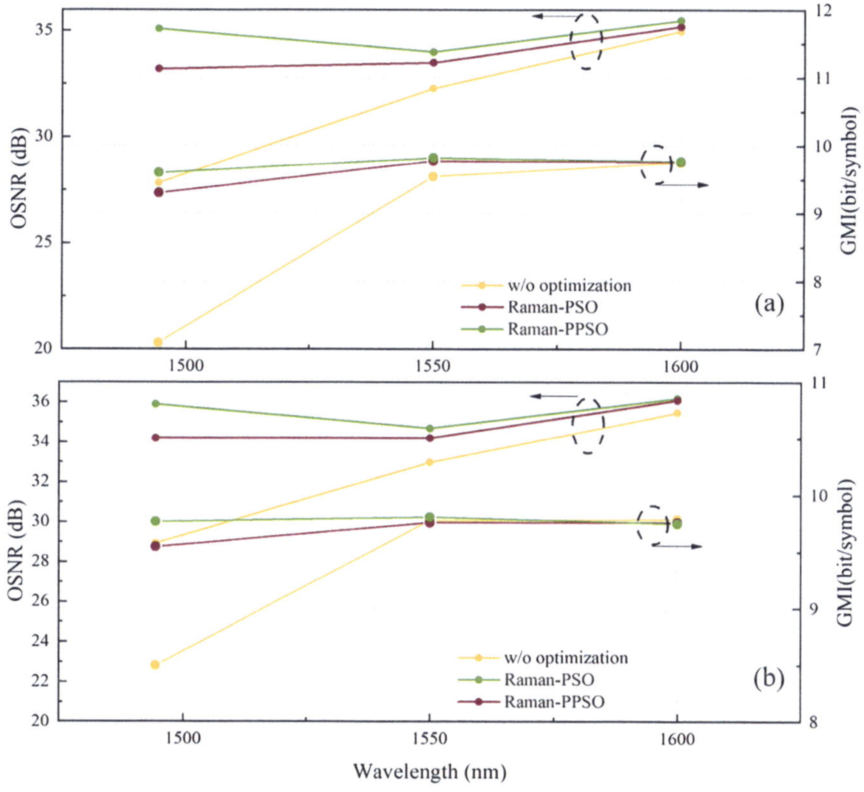

Fig. 4.18 OSNR and GMI versus wavelength for no optimization, and optimization with Raman-PSO and Raman-PPSO respectively. (**a**) Launch power = 3 dBm; (**b**) the launch power is the optimal launch power

PPSO optimization. For launch power = 3 dBm, the OSNR of the short-wavelength channel shows wavelength dependence due to the power transfer, and the OSNRs of the three channels in the S-, C-, and L-bands without optimization are 27.8, 32.3, and 35 dB. For the Raman-PPSO scheme, the OSNRs of the three channels are 34.2, 34.2, and 36.1 dB respectively. For the Raman-PPSO scheme, the OSNRs of the three channels are 35.9, 34.7, and 36.2 dB respectively. The OSNR of the short-wavelength channel is effectively enhanced. The GMI curves show a similar trend, with the GMI of the Raman-PPSO scheme being 2.5 bits/symbol higher than that of the unoptimized GMI in the S-band test channel. There is a slight improvement in the C/L band optimization. For the optimal launch power, the GMI of the Raman-PPSO scheme is 1.2 bits/symbol higher than the unoptimized GMI and 0.2 bits/symbol higher than the Raman-PSO scheme in the S-band test channel. The Raman-PSO scheme and Raman-PPSO scheme mitigate the wavelength-dependent performance degradation from the power tilt and maintain the performance of long-wavelength channels. This implies that there is a significant performance in the machine learning-assisted RA after fiber transmission in compensating for the multi-band power tilt. In addition, the PPSO algorithm is superior in terms of convergence accuracy and convergence rate compared to the PSO algorithm.

4.2.2.3 Discussion

In this section, we utilize machine learning-assisted RAs to compensate for the power tilt induced by ISRS after optical fiber transmission. A PPSO algorithm is proposed for controlling the pump wavelength and pump power of RAs. An analysis was conducted on the performance of the PPSO algorithm to compensate for the power tilt. The limitations of the Raman-PPSO scheme also need to be discussed. Specifically, the scheme is limited to the S, C, and L wavelength bands, and finding a feasible solution becomes increasingly difficult when there are fewer particles involved. The stochastic nature of the PPSO algorithm further complicates the search for a solution that can compensate for all five wavelength bands. To address these challenges, future research will focus on developing machine learning-assisted Raman amplification schemes with wider bandwidths. Additionally, there is a need to further examine the performance and cost implications of applying this scheme to multi-span systems.

4.3 Ultra-Wideband Transmission

In this subsection, we extended the available wavebands to O, E, S, C, L and U band, and constructed a 6-band optical transmission system. We demonstrated 480-channel WDM transmission of 50 GHz spaced 15 GBaud polarization division multiplexed (PDM) 256-QAM signals on this system, and the total capacity can reach 115.2 Tbit/s by using artificial neural network technology to compensate the inter-channel nonlinear interference which plays an important role limiting communication capacity.

4.3.1 System Setup

Figure 4.19a shows the setup for 6-band WDM transmission system which is established on the VPI software. The transmitter uses O, E, S, C, L and U band continuous-wave (CW) lasers with a line width of 100 kHz for transmission. In each band, 80 optical carriers with 50 GHz spaced are generated from CW lasers. The wavelength of the measured signals is set to 1304.63–1327.43 nm in the O band, 1408.78–1435.41 nm in the E band, 1456.66–1485.15 nm in the S band, 1534.92–566.58 nm in the C band, 1571.09–1604.28 nm in the L band, and 1639.79–1675.98 nm in the U band. The optical carrier is modulated by the IQ modulator which is driven by an arbitrary waveform generator (AWG). The modulated 15 GBaud PDM 256-QAM signals of each band are amplified with semiconductor optical amplifier (SOA) in the O, E and U band, thulium-doped fiber amplifiers (TDFA) in the S band, and erbium-doped fiber amplifiers (EDFA) in the C and L band. The amplified 6-band signals are multiplexed by using a multiplexer (MUX). After 20 km single-mode fiber (SMF) transmission, the WDM signal is de-multiplexed into each band using a de-multiplexer (DEMUX). The de-multiplexed signal of each band is successively amplified by SOA, TDFA,

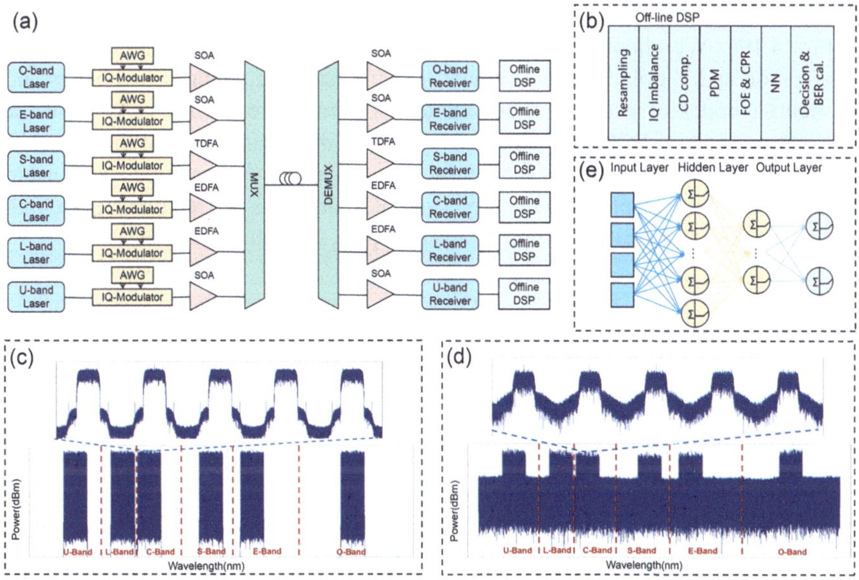

Fig. 4.19 (**a**) Experimental setup, (**b**) off-line DSP, (**c**) optical spectra before and (**d**) after transmission, (**e**) neural network structure diagram

Table 4.5 Parameters of different band

Parameter/band	O-Band	E-Band	S-Band	C-Band	L-Band	U-Band
Bandwidth range (THz)	226–230	209–213	202–206	191.5–195.5	187–191	179–183
Number of channels	80	80	80	80	80	80
Nonlinear coefficient (1/W/km)	1.55	1.44	1.39	1.32	1.29	1.23

and EDFA. After the amplification, the coherent receivers are used to detect the signals and the digitized signal is demodulated by off-line DSP as shown in Fig. 4.19 b that includes dispersion compensation, polarization de-multiplexing, frequency offset estimation, and carrier phase recovery. Then, a neural network is applied to compensate for the fiber nonlinearity. And the signal after neural network equalization is decoded and the bit error ratio (BER) is calculated. Table 4.5 shows some parameters in each band, including bandwidth range, channel number and nonlinear coefficient.

Figure 4.19 c and d show the spectrum diagram of the 6-band WDM system containing a total of 480 channels before and after 20 km SMF transmission. For the better display, the spectrum of five channels in the C-band is amplified. Wavelength multiplexing leads to the dominance of inter-channel nonlinear interference in optical fiber nonlinearity, and this interference will intensify as the number of channels increases. Due to the influence of fiber nonlinearity, the signal has been seriously damaged and distorted, which undoubtedly limits the transmission distance and quality of the signal. To mitigate the fiber nonlinear interference, we use the neural network (NN) structure based on back propagation (BP) algorithm as shown in Fig. 4.19e. We choose the in-phase (I) and quadrature (Q) components of X polarization and Y polarization as inputs. The hidden layer contains two layers and the neuron nodes of each layer are set to 300 and 10 respectively. We use the received symbols of WDM transmission system as input samples, and the transmitted symbols as target samples. The neural network learns the mapping relationship between the input samples and the target samples, and uses the relationship to train new data samples. Finally, the output layer outputs the trained samples.

4.3.2 Results and Discussion

Figure 4.20 shows the change of BER characteristics with launch power at 1550 nm wavelength in single-channel system and multi-band system with 20 channels multiplexing in each band respectively. In the single-channel system, the best BER performance (Log BER) can reach −4.063 when the launch power is −2 dBm. While in the multi-band system, the best BER performance is −3.733 at the launch power of −6 dBm. The BER performance in multi-band system is lower than that in

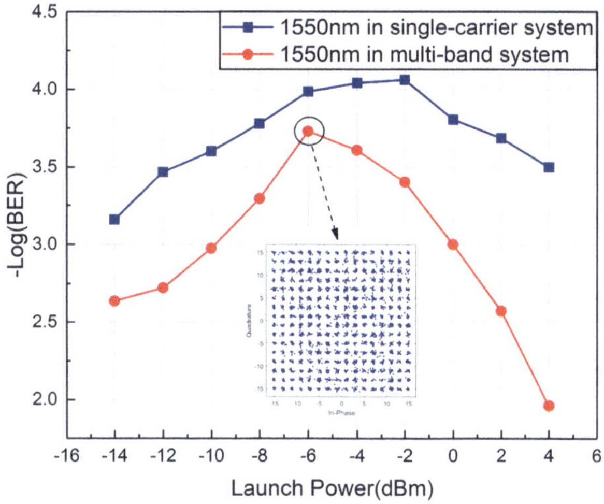

Fig. 4.20 BER versus launch power in single-channel system and multi-band system

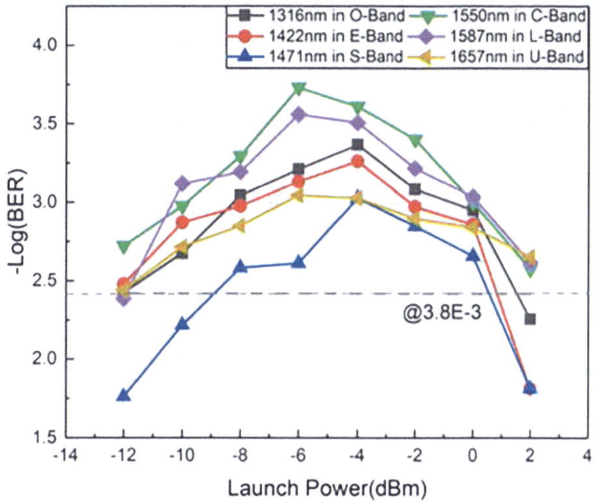

Fig. 4.21 BER versus launch power for different wavelength in multi-band system

single-channel system, which is caused by nonlinear crosstalk between channels in multi-band system.

Figure 4.21 shows the BER performance versus launch power of each band in the multi-band system at the wavelength of 1316 nm, 1422 nm, 1471 nm, 1550 nm, 1587 nm and 1657 nm, respectively. As the launch power increases, the BER performance also improves at first and reaches the optimal value at a certain launch power, then the BER performance becomes worse with the increase of launch power. This is incurred by the intensification of the fiber nonlinear effect by the increased launch power, which brings damage to the signal. The signals in O, E and S band have the best BER performance when the launch power is −4 dBm, and the signals in C, L and U band have the best BER performance at the launch power of −6 dBm.

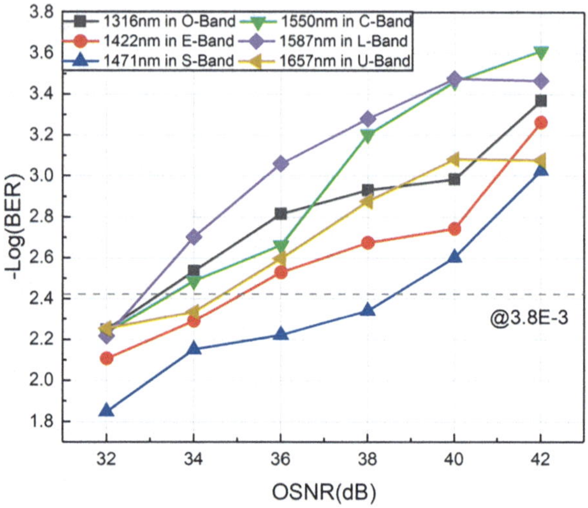

Fig. 4.22 BER versus OSNR for different wavelength in multi-band system

Although the BER of each band can reach 3.8e-3, the BER of C and L band is significantly better than that of other bands, which is caused by the characteristic difference of different bands.

We also measured the BER of each band versus optical signal-to-noise ratio (OSNR) in the multi-band system at the launch power of −4 dBm as shown in Fig. 4.22. With the increase of OSNR, the BER of each channel is decreasing accordingly. The OSNR required for each band to reach 3.8e-3 is 33.17 dB, 35.12 dB, 38.70 dB, 33.45 dB, 32.85 dB and 34.70 dB respectively and the OSNR penalty due to differences between channels is 5.53dB. Significant differences in BER between bands can be observed, and in addition to characteristic differences of bands, amplifiers may also account for it. Different amplifiers may vary slightly in gain and noise figure and this results in differences in the quality of signals.

4.4 Summary

In this chapter, we explore ultra-wideband (UWB) wavelength division multiplexing (WDM), to supercharge the capacity of coherent fiber optic communication systems. Firstly, we present a new multi-band transmission scheme using the Fast—than—Nyquist rate and a non-orthogonal technique with a duobinary filter. Then, we analyze a UWB WDM system across C + L and other bands by building it and studying channel effects. Moreover, we show an optimization plan combining a Raman amplifier and PPSO to correct spectral tilt from SRS. Finally, we achieve 480, 50GHz—spaced channels ultra-wideband transmission with PDM 256—QAM over 20 km of SSMF and reach 115.2 Tbit/s using an ANN to handle fiber nonlinearities.

References

1. Cisco, *Cisco Annual Internet Report (2018–2023) White Paper* (Cisco, San Jose, CA, 2020)
2. Cisco, *2021 Global Networking Trends Report* (Cisco, San Jose, CA, 2020)
3. D. Qian, M.F. Huang, E. Ip, Y.K. Huang, Y. Shao, J. Hu, T. Wang, 101.7 Tb/s (370 × 294-Gb/s) PDM-128QAM-OFDM transmission over 3 × 55-km SSMF using pilot-based phase noise mitigation, in *Proceedings of the Optical Fiber Communication Conference, Los Angeles, CA, USA, 6–10 March 2011*, (2011), p. PDPB5
4. B.J. Puttnam, G. Rademacher, R.S. Luís, Space-division multiplexing for optical fiber communications. Optica **8**, 1186–1203 (2021)
5. G. Rademacher, B.J. Puttnam, R.S. Luís, T.A. Eriksson, N.K. Fontaine, M. Mazur, H. Chen, R. Ryf, D.T. Neilson, P. Sillard, et al., 1.01 Peta-bit/s C+ L-band transmission over a 15-mode fiber, in *Proceedings of the 2020 European Conference on Optical Communications (ECOC), Brussels, Belgium, 6–10 December 2020*, (2020), pp. 1–4
6. G. Rademacher, B.J. Puttnam, R.S. Luís, J. Sakaguchi, W. Klaus, T.A. Eriksson, Y. Awaji, T. Hayashi, T. Nagashima, T. Nakanishi, et al., Multi-span transmission over 65 km 38-core 3-mode fiber, in *Proceedings of the 2020 European Conference on Optical Communications (ECOC), Brussels, Belgium, 6–10 December 2020*, (2020), pp. 1–4
7. D. Kong, A. Jørgensen, M. Henriksen, F. Klejs, Z. Ye, Ò. Helgason, H. Hansen, H. Hu, M. Yankov, S. Forchhammer, et al., Single dark-pulse kerr comb supporting 1.84 Pbit/s transmission over 37-core fiber, in *Proceedings of the CLEO: QELS_Fundamental Science, Washington, DC, USA, 10–15 May 2020*, (2020), p. JTh4A-7
8. J. Singh, A. Ahrens, S. Lochmann, Joint pre-and post-equalization with higher-order modulation formats in SDM-based optical MIMO systems. Photonics **9**, 876 (2022)
9. A. Ferrari, A. Napoli, J.K. Fischer, N. Costa, A. D'Amico, J. Pedro, W. Forysiak, E. Pincemin, A. Lord, A. Stavdas, et al., Assessment on the achievable throughput of multi-band ITU-T G. 652. D fiber transmission systems. J. Lightwave Technol. **38**, 4279–4291 (2020)
10. J. Renaudier, A.C. Meseguer, A. Ghazisaeidi, P. Tran, R.R. Muller, R. Brenot, A. Verdier, F. Blache, K. Mekhazni, B. Duval, et al., First 100-nm continuous-band WDM transmission system with 115Tb/s transport over 100 km using novel ultra-wideband semiconductor optical amplifiers, in *Proceedings of the 2017 European Conference on Optical Communication (ECOC), Gothenburg, Sweden, 17–21 September 2017*, (2017), pp. 1–3
11. F. Hamaoka, K. Minoguchi, T. Sasai, A. Matsushita, M. Nakamura, S. Okamoto, E. Yamazaki, Y. Kisaka, 150.3 Tb/s ultra-wideband (S-, C-, and L-bands) single-mode fibre transmission over 40-km using > 519Gb/s/A PDM-128QAM signals, in *Proceedings of the 2018 European Conference on Optical Communication (ECOC), Rome, Italy, 23–27 September 2018*, (2018), pp. 1–3
12. L. Galdino, A. Edwards, W. Yi, E. Sillekens, Y. Wakayama, T. Gerard, W.S. Pelouch, S. Barnes, T. Tsuritani, R.I. Killey, et al., Optical fibre capacity optimisation via continuous bandwidth amplification and geometric shaping. IEEE Photon. Technol. Lett. **32**, 1021–1024 (2020)
13. S. Okamoto, K. Minoguchi, F. Hamaoka, K. Horikoshi, A. Matsushita, M. Nakamura, E. Yamazaki, Y. Kisaka, A study on the effect of ultra-wide band WDM on optical transmission systems. J. Lightwave Technol. **38**, 1061–1070 (2019)
14. B.J. Puttnam, R.S. Luís, G. Rademacher, M. Mendez-Astudillio, Y. Awaji, H. Furukawa, S-, C-and L-band transmission over a 157 nm bandwidth using doped fiber and distributed Raman amplification. Opt. Express **30**, 10011–10018 (2022)
15. B.J. Puttnam, R.S. Luís, G. Rademacher, Y. Awaji, H. Furukawa, Investigation of long-haul S-, C-+ L-band transmission, in *Proceedings of the Optical Fiber Communication Conference, Washington, DC, USA, 6–11 June 2022*, (2022), p. W3C-5
16. P. Hazarika, M. Tan, A. Donodin, S. Noor, I. Phillips, P. Harper, J.S. Stone, M.J. Li, W. Forysiak, E-, S-, C-and L-band coherent transmission with a multistage discrete Raman amplifier. Opt. Express **30**, 43118–43126 (2022)

17. A. Ghazisaeidi, A. Arnould, M. Ionescu, V. Aref, H. Mardoyan, S. Etienne, M. Duval, C. Bastide, H. Bissessur, J. Renaudier, 99.35 Tb/s Ultra-wideband unrepeated transmission over 257 km using semiconductor optical amplifiers and distributed Raman amplification. J. Lightwave Technol. **40**, 7014–7019 (2022)
18. E. Virgillito, E. London, A. D'Amico, B. Correia, A. Napoli, V. Curri, Single-vs. multi-band optimized power control in C+ L WDM 400G line systems, in *Proceedings of the 2021 Optical Fiber Communications Conference and Exhibition (OFC), San Francisco, CA, USA, 6–11 June 2021*, (2021), pp. 1–3
19. I. Roberts, J.M. Kahn, J. Harley, D.W. Boertjes, Channel power optimization of WDM systems following Gaussian noise nonlinearity model in presence of stimulated Raman scattering. J. Light. Technol. **35**, 5237–5249 (2017)
20. D. Semrau, R.I. Killey, P. Bayvel, The Gaussian noise model in the presence of inter-channel stimulated Raman scattering. J. Light. Technol. **36**, 3046–3055 (2018)
21. N.A. Shevchenko, S. Nallaperuma, S.J. Savory, Maximizing the information throughput of ultra-wideband fiber-optic communication systems. Opt. Express **30**, 19320–19331 (2022)
22. C. Lasagni, P. Serena, A. Bononi, J.C. Antona, A generalized Raman scattering model for real-time SNR estimation of multi-band systems. J. Lightwave Technol. **41**(11), 3407–3416 (2023) early access
23. F. Hamaoka, M. Nakamura, S. Okamoto, K. Minoguchi, T. Sasai, A. Matsushita, E. Yamazaki, Y. Kisaka, Ultra-wideband WDM transmission in S-, C-, and L-bands using signal power optimization scheme. J. Lightwave Technol. **37**, 1764–1771 (2019)
24. B. Correia, R. Sadeghi, E. Virgillito, A. Napoli, N. Costa, J. Pedro, V. Curri, Optical power control strategies for optimized C+ L+ S-bands network performance, in *Optical Fiber Communication Conference*, (Optica Publishing Group, Washington, DC, 2021), p. W1F-8
25. B. Correia, R. Sadeghi, E. Virgillito, A. Napoli, N. Costa, J. Pedro, V. Curri, Power control strategies and network performance assessment for C+ L+ S multiband optical transport. J. Opt. Commun. Netw. **13**, 147–157 (2021)
26. H. Luo, J. Lu, Z. Huang, C. Yu, C. Lu, Optimization strategy of power control for C+ L+ S band transmission using a simulated annealing algorithm. Opt. Express **30**, 664–675 (2022)
27. S.E. Landero, I.F. de Jauregui Ruiz, A. Ferrari, D. Le Gac, Y. Frignac, G. Charlet, Link power optimization for S+ C+ L multi-band WDM coherent transmission systems, in *Proceedings of the 2022 Optical Fiber Communications Conference and Exhibition (OFC), San Diego, CA, USA, 24–28 March 2022*, (2022), pp. 1–3
28. U.C. De Moura, M.A. Iqbal, M. Kamalian, L. Krzczanowicz, F. Da Ros, A.M.R. Brusin, A. Carena, W. Forysiak, S. Turitsyn, D. Zibar, Multi–band programmable gain Raman amplifier. J. Lightwave Technol. **39**, 429–438 (2020)
29. A. Napoli et al., Perspectives of multi-band optical communication systems, in *Proceedings of the 23rd Opto-Electronics and Communications Conference*, (2018), pp. 1–2
30. M. Jinno et al., Design of ultra-wide band (1450–1650 nm) WDM transmission systems considering intra-and inter-wavelength-band nonlinear interactions, in *Proceedings of the European Conference on Networks and Optical Communications (ECOC)*, vol. 3, (1998), pp. 205–208
31. R. Freund, L. Molle, F. Raub, C. Caspar, M. Karkri, C. Weber, Triple-(S/C/L)-band WDM transmission using erbium-doped fibre amplifiers, in *Proceedings of the 31st European Conference Optical Communication, Glasgow, 2005, Paper Mo.4.2.3*, (2005)
32. G. Forney, Maximum-likelihood sequence estimation of digital sequences in the presence of inter symbol interference. IEEE Trans. Inform. Theory **18**(3), 363–337 (1972)
33. J. Hayes, The Viterbi algorithm applied to digital data transmission. Commun. Soc. **13**(2), 15–20 (1975)
34. M.S. Baek, N.H. Hur, H. Lim, Novel interference cancellation technique based on matrix computation for FTN communication system, in *Proceedings of IEEE Military Communications Conference, Baltimore*, (2014), pp. 830–834

35. G.P. Agrawal, *Lightwave Technology: Telecommunication Systems* (John Wiley & Sons, Hoboken, NJ, 2005)
36. A. Carena, V. Curri, G. Bosco, P. Poggiolini, F. Forghieri, Modeling of the impact of nonlinear propagation effects in uncompensated optical coherent transmission links. J. Lightwave Technol. **30**, 1524–1539 (2012)
37. D. Semrau, E. Sillekens, P. Bayvel, R.I. Killey, Modeling and mitigation of fiber nonlinearity in wideband optical signal transmission. J. Opt. Commun. Netw. **12**, C68–C76 (2020)
38. K. Minoguchi, S. Okamoto, F. Hamaoka, A. Matsushita, M. Nakamura, E. Yamazaki, Y. Kisaka, Experiments on stimulated Raman scattering in S-and L-bands 16-QAM signals for ultra-wideband coherent WDM systems, in *Proceedings of the Optical Fiber Communication Conference, San Diego, CA, USA, 11–15 March 2018*, (2018), p. Th1C-4
39. G. Saavedra, D. Semrau, M. Tan, A. Iqbal, D.J. Elson, L. Galdino, P. Harper, R.I. Killey, P. Bayvel, Inter-channel stimulated Raman scattering and its impact in wideband transmission systems, in *Proceedings of the 2018 Optical Fiber Communications Conference and Exposition (OFC), San Diego, CA, USA, 11–15 March 2018*, (2018), pp. 1–3

Chapter 5
Space Division Multiplexing (SDM) Technology

In the situation where other dimensions of fiber optic transmission cannot be broken through, space division multiplexing technology provides a new development direction for improving the capacity of fiber optic transmission systems, which will increase the transmission capacity of the system in an order of magnitude. Space division multiplexing technology achieves several times the carrying capacity by increasing the number of fiber cores or spatial patterns within the same optical fiber.

Space division multiplexing technology, theoretically through multi-dimensional multiplexing of space and mode, can increase the transmission capacity of a single fiber by tens of times on the original basis, and is expected to solve the future data transmission capacity crisis. It is where the development of ultra large capacity fiber optic technology is trending. Space division multiplexing technology treads on the heels of the development of fiber optic communication, as a major innovation in the field of fiber optic communication technology, and it indicates where the next generation of optical communication technology is heading.

5.1 Basic Principle

The basic concept of space division multiplexing technology is to divide space into multiple independent channels and transmit different signals to different channels, thereby improving channel utilization. This method can process multiple signals simultaneously, improve transmission capacity and performance, and thus achieve efficient communication.

In fiber optic communication, space division multiplexing technology can currently be divided into the following three channel processing methods: multicore fiber technology (Fig. 5.1), few mode fiber technology, and multi core few mode fiber technology. Multicore fiber technology refers to the presence of multiple independent fiber cores in a common cladding area, which can simultaneously transmit signals from multiple channels. Few mode fiber technology is a type of

F. Tian, X. Xin, *High-Speed Large Capacity Optical Fiber Communications*, Wireless Networks, https://doi.org/10.1007/978-3-031-94211-2_5

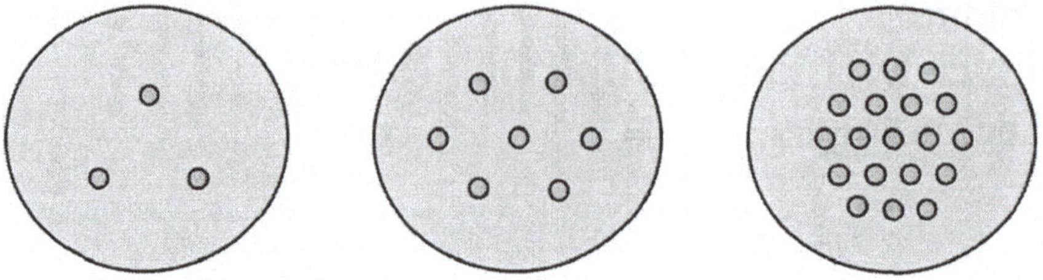

Fig. 5.1 Multi-core fiber cross section structure

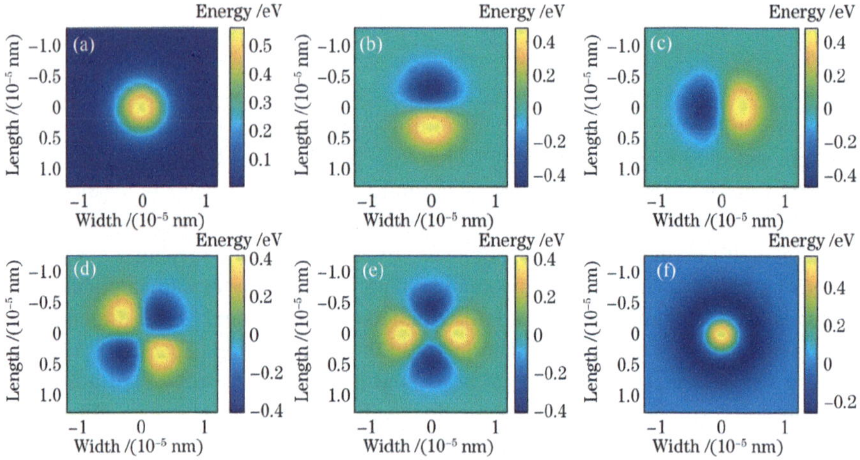

Fig. 5.2 Mode field diagrams for six modes. (**a**) LP_{01}; (**b**) LP_{11a}; (**c**) LP_{11b}; (**d**) LP_{21a}; (**e**) LP_{21b}; (**f**) LP_{02}

fiber with a sufficient core area to utilize several independent spatial modes to transmit parallel data streams. Typically, the capacity of a few mode fiber is proportional to the number of modes. Few mode multi-core fiber technology is a combination of multi-core fibers and few mode fibers. Multiple optical modes are transmitted in multiple cores of a single fiber, which can significantly improve the transmission rate and the capacity of the system. It is a major development trend with good development prospects in optical space division multiplexing technology.

The six modes of the six-mode fiber we use are LP_{01}, LP_{11a}, LP_{11b}, LP_{21a}, LP_{21b}, LP_{02}, where LP_{11a} and LP_{11b}, LP_{21a} and LP_{21b} are two pairs of degenerate modes, and the mode field of each mode in the fiber is $E(r, \psi)$, where r is the polar radial quantity and ψ is the polar Angle [1]. The mode fields of different modes are shown in Fig. 5.2.

Mode crosstalk occurs in low-mode fibers under the influence of various factors such as temperature, pressure, bending, etc. Here, various factors are equivalent to the displacement and rotation of the fiber segment [2], as shown in Fig. 5.3, where the left circle represents the fiber segment m, the right circle represents the fiber

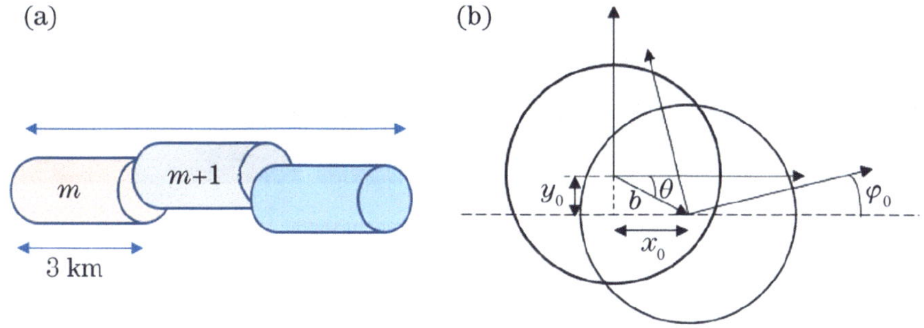

Fig. 5.3 schematic of displacement and rotation of optical fiber segment. (**a**) Optical fiber segments; (**b**) displacement and rotation of cross section of optical fiber

segment $m + 1$, b and θ are the displacement distance and direction of the fiber segment, and φ_0 is the rotation angle of the fiber segment. x_0 and y_0 are the lateral and vertical displacement distances, respectively.

K_{ij} represents the crosstalk magnitude from the j mode of the m segment fiber to the i mode of the $m + 1$ segment fiber. The calculation method is the reintegral normalization of the mode field function of the two modes [1], and the expression is

$$K_{ij} = \frac{\iint E_j^m(\boldsymbol{r}, \psi) \cdot E_i^{m+1}(\boldsymbol{r} - \boldsymbol{b}, \psi - \varphi_0)}{\sqrt{\iint \left[E_j^m(\boldsymbol{r}, \psi)\right]^2 \cdot \iint \left[E_i^{m+1}(\boldsymbol{r} - \boldsymbol{b}, \psi - \varphi_0)\right]^2}} \tag{5.1}$$

Where: b is the displacement vector. The mode field of the fiber in section m is unchanged, and the mode field of the fiber in section $m + 1$ is similarly shifted and rotated due to the displacement and rotation of the fiber segment. By substituting all mode fields into Eq. (5.1), all mode crosstalk K_{ij} of six-mode fiber is calculated.

The total mode crosstalk is expressed as $\boldsymbol{H}$, and the expression is

$$\boldsymbol{H} = \begin{bmatrix} K_{11} & K_{12} & K_{13} & K_{14} & K_{15} & K_{16} \\ K_{21} & K_{22} & K_{23} & K_{24} & K_{25} & K_{26} \\ K_{31} & K_{32} & K_{33} & K_{34} & K_{35} & K_{36} \\ K_{41} & K_{42} & K_{43} & K_{44} & K_{45} & K_{46} \\ K_{51} & K_{52} & K_{53} & K_{54} & K_{55} & K_{56} \\ K_{61} & K_{62} & K_{63} & K_{64} & K_{65} & K_{66} \end{bmatrix} \tag{5.2}$$

Signal transmission speed of different modes is different, and the time difference to reach the receiving end is differential mode delay [3].

Combining mode crosstalk and differential mode delay, the channel function $\boldsymbol{T}$ of the few-mode fiber can be expressed by $\boldsymbol{D}$ and $\boldsymbol{H}$. Suppose that the sent signal of six channels is represented as $\boldsymbol{X}$ by row arrangement, the received signal is represented as $\boldsymbol{Y}$ by row arrangement, and the channel function is $\boldsymbol{T}$,

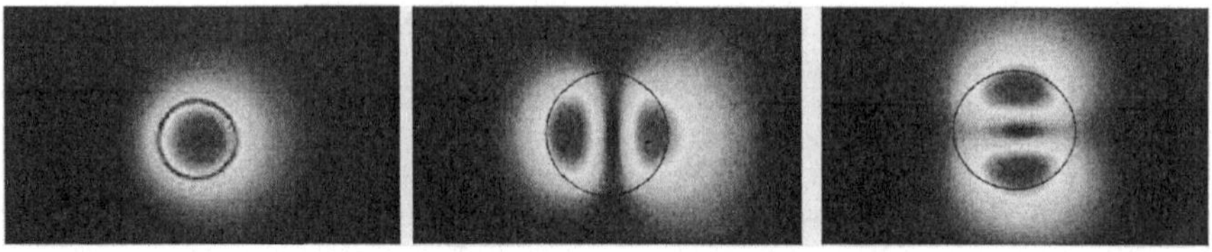

Fig. 5.4 Mode field deviation leakage of optical fiber in bent state

$$Y = F^{-1}[\boldsymbol{T} \cdot F(\boldsymbol{X})] \tag{5.3}$$

Where: F^{-1} and F are inverse Fourier transform and Fourier transform respectively.

Based on the model coupling theory and power coupling theory, the power and coupling coefficient of crosstalk are obtained. It can be found that when it is a homogeneous multi-core fiber and the bending radius is relatively small, the power coupling coefficient and mode coupling coefficient meet the following relations:

$$XT = 2\frac{k^2}{\beta}\frac{R}{D_{mn}}L = h_{mn}L \tag{5.4}$$

From formula (5.5), we can calculate the mode coupling coefficient k_{mn} of the two adjacent cores, where the denominator represents the power of the fiber end face, and the numerator is expressed as the overlapping integral of the electric field of the waveguide m and the electric field generated by the waveguide n in the waveguide m. Furthermore, intercore crosstalk can be calculated in combination with the above.

$$k_{mn} = \frac{\omega\varepsilon_0 \iint (N^2 - N_n^2)E_m^* E_n dxdy}{\iint E_m^* H_m + E_m H_m^* dxdy} \tag{5.5}$$

In the analysis of intercore crosstalk of multi-core fiber, we introduce the random factor, which is caused by the torsion or bending that inevitably occurs in practical use.

Optical fiber will inevitably bend and twist during actual use. In this case, the transmission characteristics inside the optical fiber will change. After bending, the mode field inside the core will leak outwardly, resulting in bending loss, as shown in Fig. 5.4, which shows the distribution of mode field in the bending state of the optical fiber obtained in the experimental simulation. The mode field distribution between the core is tilted to one side, and in practical applications, energy will leak to one side, resulting in loss. In multi-core fiber, because of the large number of core cores inside the cladding, the cladding between the core becomes thinner, and the bending loss is further aggravated.

According to perturbation theory, the bending loss calculation method applicable to both single-clad fiber and multi-clad fiber is obtained:

$$L_S = 10\,log_{10}(exp(2\alpha L)) \tag{5.6}$$

$$2\alpha = -2\frac{\kappa^2}{2\pi\beta V^2 K_1^2(\alpha\gamma)} Im\left(\int_{-\infty}^{\infty} H_1(\zeta)A_i[X_2(0,\zeta)]d\zeta\right) \tag{5.7}$$

The effective mode field area represents the power distribution of the core mode field in the fiber, and represents the optical power density at the end face. The effective area of the fiber directly affects the nonlinear effect in the fiber and is an important parameter to describe the performance of the fiber. The larger the value, the smaller the nonlinear effect. The effective area is also an important performance index of fiber. Generally speaking, when optical signals are transmitted in the fiber, a small amount of mode field will leak into the cladding. In this case, it is not accurate to calculate the mode field area directly through the core diameter of the fiber, and it is necessary to integrate the electric field intensity E at the entire end of the fiber. The specific expression is shown as Eq. (5.8): Where A_{eff} represents the effective mode field area, E_t represents the transverse electric field strength of the mode electric field, and S represents the entire end area of the fiber.

$$A_{eff} = \frac{\left(\int\int_S |E_t|^2 dxdy\right)^2}{\int\int_S |E_t|^4 dxdy} \tag{5.8}$$

5.2 SDM Transmission System

In recent years, as the theoretical limit of the single-mode Fiber (SMF) is approached, the space division multiplexing (SDM) transmission has become a promising candidate technology to break through the Shannon limit of the conventional transmission systems [4], whose transmission medium is generally performed over multi-core Fiber (MCF), few-mode Fiber (FMF) or the combinations between them. It could offer spatially parallel channels in a single fiber. However, there are more transmission impairments such as mode-dependent loss (MDL), crosstalk between different cores, dispersion and even with distortions existed in SMF, which unfavorably influenced the transmission performance [5, 6].

Up till now, multi-input and multi-output (MIMO) equalization is the most effective way to compensate SDM systems' impairments [7]. MIMO usually uses blind equalization or non-blind equalization in optical research, compared with the former, the latter has faster convergence speed and stronger stability. The least mean squares (LMS) algorithm is considered as an attractive method in SDM MIMO system because it strikes a good balance among performance, hardware complexity and dynamic speed [8]. The adaptive LMS algorithm can be implemented in time domain or frequency domain. Frequency domain method is more efficient compared with time domain method [9–11].

5.2.1 Training Sequence Aided Constant Modulus Algorithm (TA-CMA)

In recent years, data traffic has exploded thanks to the rise of cloud services and high definition (HD) Video. Intensity modulation/direct-detection (IM/DD) technology based on pulse amplitude modulation (PAM) has received a lot of attention with a simple and low-cost scheme for short-reach application [12–14]. Meanwhile, probabilistic shaping (PS) technology is an efficient transmission technology with sound anti-noise performance. Flexible spectral efficiency (SE) or close to the Shannon limit can be achieved by adaptively changing the signal probability distribution according to the signal-to-noise ratio (SNR). The application of PS technology in IM/DD systems is reported to be gradually increasing [15–18].

On the other hand, compared with the traditional single-mode fiber, the space division multiplexing (SDM) technology based on multi-core fiber (MCF) and few-mode fiber (FMF) has greatly increased the capacity limit [19, 20].

We experimentally investigate the performance of uniform and probabilistically shaped 8-level pulse amplitude modulation signals over 10-km 6-mode gradient index fiber. A training sequence aided constant modulus algorithm (TA-CMA) is proposed to equalize the signal impairments from the FMF channel. The experimental results indicate that the proposed TA-CMA can achieve a ~1 dB receiver power gain at 7% overhead FEC threshold (3.8×10^{-3}).

Typical constant modulus algorithm (CMA), a common blind equalization algorithm, is effective for constant modulus modulation. The cost function of CMA is given by [21]:

$$J_{CMA}(n) = E\left[\left(|\widetilde{x}(n)|^2 - K_{CMA}\right)^2\right] \tag{5.9}$$

Where $\widetilde{x}(n)$ represents equalizer output, K_{CMA} is a constant, and E[.] denotes the expected value.

The weight update function of the taps is as follows

$$w(n+1) = w(n) + \mu\widetilde{x}(n)\left[K_{CMA} - |\widetilde{x}(n)|^2\right]y^*(n) \tag{5.10}$$

where w(n) is the weight of equalizer taps, μ is the step size of the equalizer, which denotes any small real value, and y(n) is the input of equalizer. y*(n) is the complex conjugate of the received symbols. Assuming the length of taps is L, the output of the equalizer is given by:

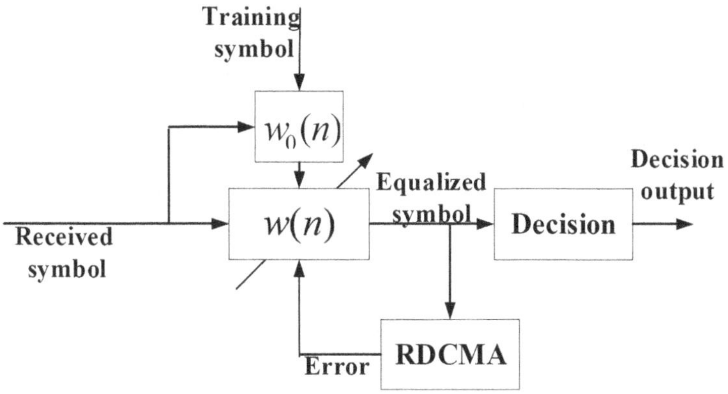

Fig. 5.5 the schematic of the training sequence aided constant modulus algorithm

$$\tilde{x}(n) = \sum_{i=0}^{L-1} w_i(n) y(n-i) \tag{5.11}$$

This algorithm is optimal for constant modulus signals because all the signal points are on a ring. The signal points of non-modulus mode signals are not on the same ring, so different moduli are needed to equalize the signals on different rings. There are several cost functions given by radius-directed (RD) method in CMA algorithm for signals with multiple moduli [22]:

$$J_{\mathrm{RDCMA}}(n) = E\left[\left(|\tilde{x}(n)|^2 - K_{\mathrm{RDCMA}}(j)\right)^2\right]_{j=1,2.....M} \tag{5.12}$$

where K_{RDCMA} is a real value determined by modulus of the nearest constellation symbol and M is the number of moduli.

In this section, the TA-CMA is proposed to overcome the dynamic characteristics and mode coupling in the few-mode links, which can improve the accuracy of error convergence. Figure 5.5 illustrates the schematic of TA-CMA.

Figure 5.5 shows the schematic diagram of the experimental setup. At the transmitter, the light is emitted from a tunable laser at 1550 nm with 100 kHz linewidth. The laser output power is fixed to 13 dBm. A PRBS is used to map onto PS-PAM-8 constellation by CCDM and then is sent to the AWG operated at 60Gsa/s. Afterwards, the electrical signal is modulated by a 40 GHz MZM with a 20GBaud symbol rate. The output optical power is 4.1 dBm due to the insertion loss of the MZM. The modulated optical signal is amplified by an EDFA to provide the power gain into the 1 × 6 OC. Then, the light is rotated by a PC and coupled into FMF with a photonic lantern 6-mode multiplexer.

After the transmission of 10 km FMF, a demultiplexer is used to decouple the signal. Due to attenuation and coupling in the mode multiplexer and fiber, the optical output power of LP01 is 4.2 dBm, while that of LP11a is 3.1 dBm. An EDFA and a VOA is used to control the received power. The transmitted signal is detected by a

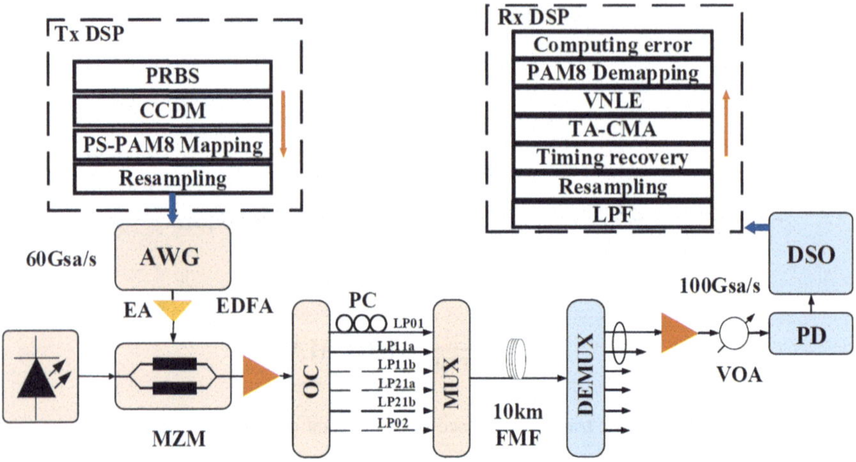

Fig. 5.6 Experimental setup for the 20GBaud PS-PAM-8 system. *PRBS* pseudo-random binary sequence, *CCDM* constant composition distribution matcher, *AWG* arbitrary waveform generator, *EA* electric amplifier, *MZM* Mach-Zehnder modulator, *OC* optical coupler, *PC* polarization controller, *MUX* multiplexer, *FMF* few-mode fiber, *DEMUX* demultiplexer, *EDFA* Erbium-doped fiber amplifier, *VOA* variable optical attenuator, *PD* photodiode, *DSO* digital storage oscilloscope, *LPF* low pass filter, *TACMA* training sequence aided constant modulus algorithm, *VNLE* Volterra nonlinear equalization

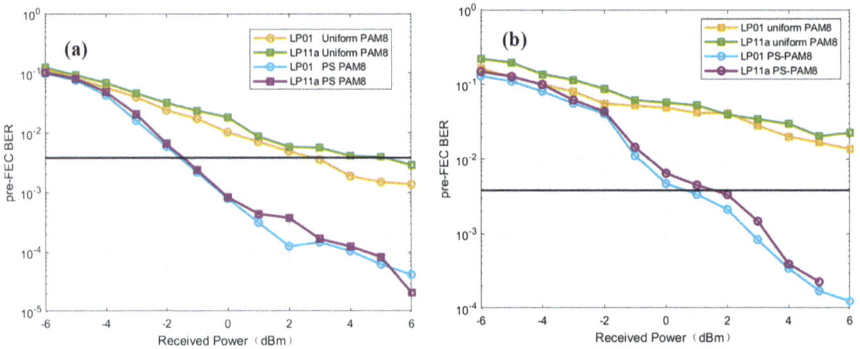

Fig. 5.7 Experimental results of PAM-8 with B2B (**a**) and after 10 km FMF transmission (**b**) for uniform and PS signal

50 GHz PD. Then a DSO with 100GSa/s and 20 GHz analog bandwidth is used to capture the signal. The offline DSP is performed at the receiver which includes LPF, time recovery, TA-CMA, three-order VNLE and the BER calculation (Fig. 5.6).

Figure 5.7 shows the experimental results of the optical back-to-back (B2B) and after a transmission of 10 km as pre-FEC BER of LP01 and LP11a modes versus received power. The performance of PS-PAM-8 signal is improved apparently with a fixed entropy (2.34-bit/symbol) employing Maxwell-Boltzmann (MB) distribution. From Fig. 5.7 a for B2B scenarios, it can be seen that the

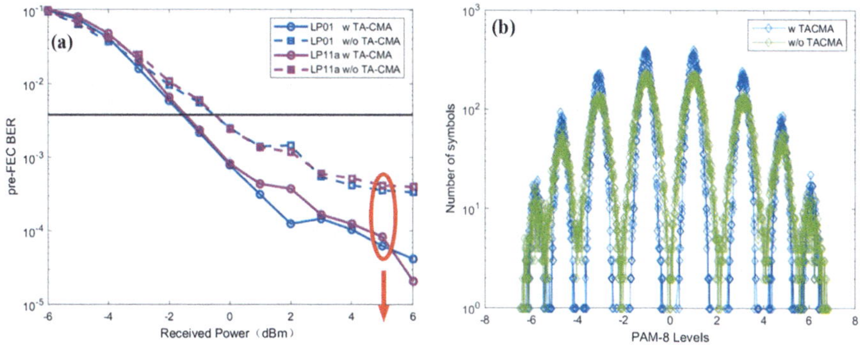

Fig. 5.8 (**a**) Experimental results of the experiment with or without TACMA in the B2B scenarios, (**b**) statistical graph of received signal with and without TA-CMA

sensitivity of the receiver in two modes is improved by more than 1 dB at 7% HD-FEC overhead (3.8×10^{-3}). After a transmission of 10 km, the BER of the uniform PAM8 signal is reduced to 1.0×10^{-2} while the received power is 6 dBm. The performance of uniform PAM-8 is higher than FEC threshold within the range of received power we measured while that of PS-PAM-8 was lower than FEC threshold at 2 dBm. It is noted that the PS gain after 10 km FMF transmission is larger compared with B2B scenarios. Considering training, PS and 7% HD-FEC overhead, the net bit rate of PS-PAM-8 is 43.5 Gbit/s.

Figure 5.8 shows the results of the experiment with or without TA-CMA in the B2B scenarios. The received power gain using TA-CMA is about 1 dB for both two modes. While the received power is 5 dBm, the pre-FEC BER of PSPAM-8 signals on LP01 and LP11a can be reduced to 1.74×10^{-4} and 2.31×10^{-4} respectively. As illustrated in Fig. 5.8 b, compared without TA-CMA, when the received power is 3 dBm, the signal amplitude with TA-CMA is more convergent.

In this section, the performance of 20GBaud uniform PAM-8 and PS-PAM-8 signals are experimentally demonstrated with pre-FEC BER performance over 10-km FMF (LP01/LP11a) based on a proposed training sequence aided CMA. The experimental results indicate that TA-CMA can achieve a ~1 dB gain for receiver sensitivity at 7% overhead HD-FEC threshold and the net bit rate of PS-PAM-8 is 43.5 Gbit/s.

5.2.2 *Principal of VSS-FDLMS*

In the frequency domain, the input signal is divided into N points in each block and for the tap coefficient, it is updated every N points by error signals. Considering the LMS in the frequency domain, it is the best choice that the order L of the adaptive filter is equal to the block length N in terms of computational complexity. In this way, it can not only ensure the same convergence speed as the time-domain LMS algorithm, but can also use the fast FFT technology to calculate the linear convolution with the cyclic convolution of the sequence, thus greatly reducing the amount of calculation. Generally, there are two methods for fast convolution operation, overlapping storage and overlapping addition, which are using discrete Fourier transform to calculate linear convolution. The overlapping storage method is the more commonly used method of the two, and its efficiency is the highest when 50% overlap.

The input in $i - th$ block is expressed as $x(i) = [x(iN), x(iN+1), \ldots x(iN+N-1)]$, and the weight coefficient is $w(i) = [w_0(i), w_1(i), \ldots w_{N-1}(i)]$. The output length of frequency domain filtering is also N, and the $i - th$ block of the output $y(n)$ is

$$y(i) = x^T(i)w(i) \tag{5.13}$$

For each block i, the 2 N sequence is composed of $x(i - 1)$ and $x(i)$. When i = 1, N zeros are added in front of $x(1)$ and after the corresponding $w(i)$.

Then use FFT transform to complete the filter coefficient adaptation in the frequency domain.

$$X(i) = fft[x(iN-N), \ldots x(iN-1), x(iN), x(iN+1), \ldots x(iN+N-1)] \tag{5.14}$$

$$W(i) = fft\left[w^T(i)0, 0, \ldots 0\right] \tag{5.15}$$

The output of the adaptive filter is the time-domain sequence without the first N zeros, which can be expressed as

$$y(n) = \{ifft[W(i)X(i)]\}_{last\ N\ elements} \tag{5.16}$$

According to the frequency domain weight coefficient recurrence equation and the principle of the steepest descent method, the weight update function of the taps is as follows.

$$W(i+1) = W(i) + \mu\nabla(i) \tag{5.17}$$

$$\nabla(i) = \begin{bmatrix} \{ifft[W(i)X(i)]\}_{last\ N\ elements} \\ 0 \end{bmatrix} \tag{5.18}$$

$$E(i) = fft\begin{bmatrix} 0 \\ e(i) \end{bmatrix} \tag{5.19}$$

$$d(i) = [d(iN), d(iN+1), \ldots d(iN-1)]^T \tag{5.20}$$

$$e(i) = [e(iN), e(iN+1), \ldots e(iN-1)]^T = d(i) - y(i) \tag{5.21}$$

Where $e(i)$ is the error vector, $d(i)$ is the expected vector, and $E(i)$ is the frequency domain form of the error vector.

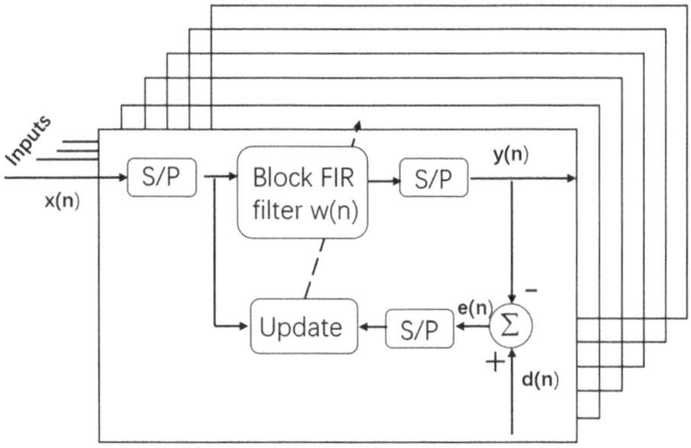

Fig. 5.9 The schematic of the frequency domain LMS algorithm in SDM

For the frequency domain LMS divides the data into blocks and filters them block by block before updating the tap, and the variable step size will be used to balance the power changes between the blocks, so that the convergence speed of the algorithm is basically the same (Fig. 5.9).

The step size is expressed as

$$u_k = \mu_0 / P_k \tag{5.22}$$

μ_0 is the initial step size, P_k represents the estimated average power of the $k - th$ block.

$$P_k(i) = \alpha P_k(i-1) + (1-\alpha)|e(i)|^2 \tag{5.23}$$

$0 < \alpha < 1$, and its function is to control the speed of previous iterations.

Figure 5.10 shows the difference between the proposed variable step size FD-LMS algorithm and the normal fixed step size FD-LMS algorithm.

5.2.3 *Simulation Setup and Results*

Figure 5.11 shows the schematic diagram of the simulation setup. In order to test the transmission performance of the system, we built a 7-core 6-mode SDM transmission system on the VPI platform, which includes 42 channel dual-polarization QAM signal generator, ideal SDM multiplexing and demultiplexing device, 42 channel polarization-diversion coherent receivers (PD-CRXs), and off-line DSP processing.

At the transmitter, a pseudo-random binary sequence (PRBS) is sent to the arbitrary waveform generator (AWG). The light emitted from a tunable laser at 1550 nm with 100 kHz linewidth is injected into a polarization-diversion I/Q modulator. The 42 PDM-16QAM signals are independent of each other to reduce the correlation between channels.

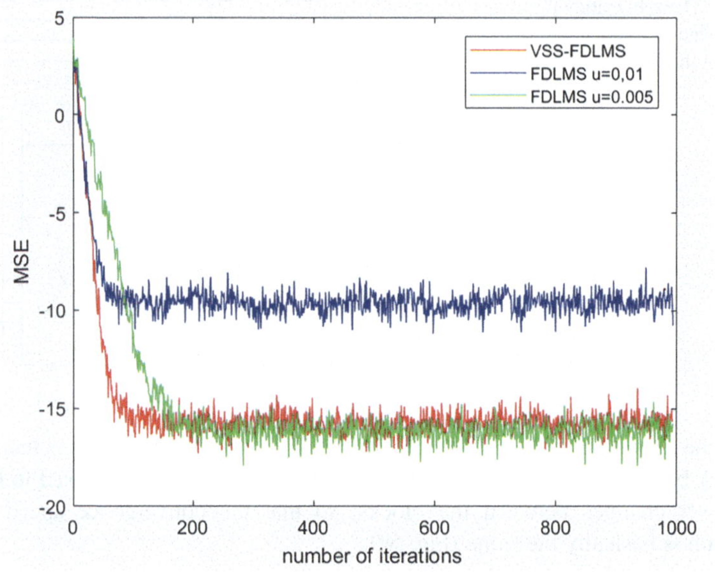

Fig. 5.10 Comparison between variable step size and fixed step size iterations

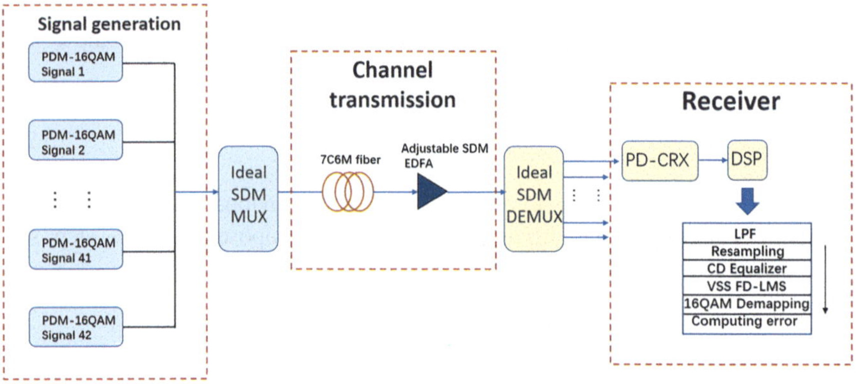

Fig. 5.11 Simulation setup for the 20GBaud PDM-16QAM SDM system. *EDFA* Erbium-doped fiber amplifier, *PD-CRX* polarization-diversion coherent receiver, *MUX* multiplexer, *7C6M* 7 core-6 mode, *DEMUX* demultiplexer, *DSP* digital signal processing, *LPF* low pass filter, VSSFDLMS: variable-step-size FD-LMS

In the transmission link, through an ideal SDM MUX, 42 optical signals are coupled into a 7-core 6-mode fiber and then transmitted through an ideal tunable SDM amplifier, whose noise figures (NFs) are fixed at 4 dB. The dispersion of the fiber is 20 ps /nm*km and differential group delay (DGD) of modes LP01, LP11, LP21, LP02 are set at 0, 150, 200, 250 ps/km. The attenuation of fiber is 0.2 dB/km.

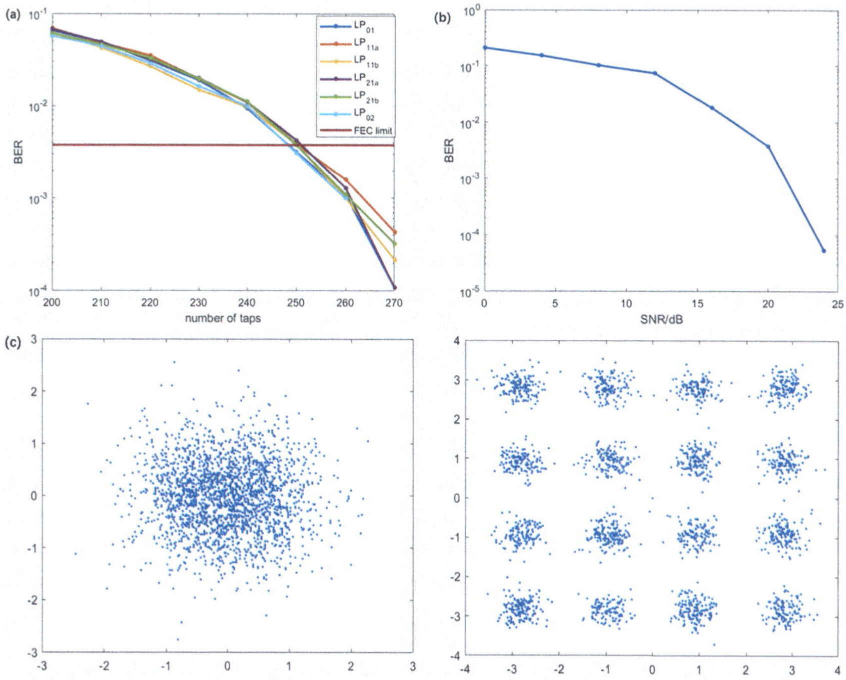

Fig. 5.12 BER performance of (**a**) different taps at SNR = 20 dB, (**b**) different SNR at taps number k = 270. (**c**) Constellation diagram before and after the equalization

At the receiver, the signals out of the ideal SDM DEMUX are detected by PD-CRXs. The offline DSP includes LPF, resampling, CD equalizer, VSS-FDLMS and the BER calculation.

We took the third core as an example. Figure 5.12 a shows the BER curves of the 6 modes of the third core after transmission of 100 km. It can be seen that when the number of taps is up to more than 250, BER will be less than 3.8×10^{-3}, which is 7% HD-FEC overhead (SNR = 20 dB). Therefore, we choose 270 as the number of taps for the transmission because more taps will affect complexity and transmit performance. Figure 5.12 b shows the relationship between BER and SNR when the number of taps is fixed. Figure 5.12 c shows the constellation diagram changes before and after compensation with SNR = 20, k = 270.

5.3 Experimental with XT-MIMO-Volterra Algorithm Over a 7-Core Fiber with a KK Receiver

With the gaining popularity of 5G mobile communication and the promotion of the shift to online services in recent years, the demand for data centers and optical access networks with higher capacity has increased dramatically [23–25]. In the context of traditional single-mode fiber capacity gradually approaching the Shannon limit, space division multiplexing (SDM) is gradually becoming a research hotspot and inevitable trend in the future development of optical communications [26–34]. Uncoupled multicore fiber (UC-MCF) is a promising alternative to various SDM transmission media, and some high-capacity short- and medium-haul optical transmission and optical interconnections have been widely reported to increase the capacity of fiber-optic transmission systems [35, 36]. It is demonstrated that a transmission capacity of 1.05 Pbit/s is obtained over a distance of 3 km using a 14-core hybrid multicore fiber in [37], transmitting 32QAMOFDM in the single-mode core and QPSK in the few-mode core. The 37-core fiber is also used for short distance transmission, supporting 64- and 256-QAM signals with a capacity of 1.84 Pbit/s after 7.9 km transmission [38]. Multicore fibers offer significant potential for applications in short- and medium-range optical transmission thanks to their increased transmission bandwidth and interconnect density [39].

However, since the multiple receivers are required in multicore optical transmission systems, the costs will significantly increase. Especially, multiple coherent receivers are unable to meet the short- and medium-range transmission and optical interconnection requirements due to their high costs, excessive power consumption, and large module size [40]. Meanwhile, single-sideband (SSB) modulation techniques based on intensity modulation and intensity detection (IM/DD) systems cannot overcome the signal impairment caused by frequency selective fading due to fiber chromatic dispersion [41]. The Kramers–Kronig (KK) receiver has been proposed in recent years as a potentially powerful technology to solve these problems [42, 43]. The KK scheme eliminates signal-to-signal beat interference (SSBI) and overcomes the limitation that a single optical detector can only detect and process the intensity information of the optical signal. It reduces the carrier signal power ratio (CSPR) requirements and allows for the reconstruction of phase information from the light intensity by using a single optical detector, provided that the signal meets the minimum phase conditions [44]. It has a bright future in short- and medium-range transmission scenarios using multicore fiber optic transmission systems owing to its low cost and superior performance.

Meanwhile, a non-negligible problem with UC-MCF in the transmission is the inter-core crosstalk (XT) caused by field coupling between cores and the losses in each core. Recent researches for UC-MCF have shown that XT can ultimately limit transmission performance and system capacity when crosstalk is minimized in the design of multicore fibers [45]. The introduction of higher-order modulation signals into MCF transmissions can immensely improve spectral efficiency, yet they are more susceptible to crosstalk. Moreover, the increasing number of cores also causes serious crosstalk. In SDM systems, the multiple-input multiple-output (MIMO) equalization algorithm is generally required to eliminate the effects of XT. For instance, an all-optical MIMO photonic integrated circuit in a mode division multiplexing (MDM) system is demonstrated in [46], successfully demultiplexing six modes after 30 km of few-mode fiber transmission. A real-time 6 × 6 MIMO-DSP is used in [47] to resist crosstalk in coupled-core multicore fiber (CC-MCF) transmission. The real-value 8 × 2 MIMO-DSP is also used for real-time transmission in a 48 km 10-mode MDM system [48]. The use of MIMO-DSP for crosstalk cancellation in UCMCF transmission systems is demonstrated in [45, 49]. The SDM transmission experiments typically employ linear MIMO equalization techniques to demultiplex spatial channels, while nonlinear MIMO schemes typically have superior performance [50]. For example, MIMO algorithms based on the Volterra series can be used to compensate the transceiver nonlinearity in SDM transmission system.

In this section, an experimental demonstration of a 16-QAM signal with a net rate of 261.7 Gbit/s transmitted over a 2.5 km seven-core fiber with a KK receiver is investigated. The relationship between CSPR, signal power, frequency spacing, and received power is investigated in this seven-core fiber transmission system, and a point is identified where a balance is achieved between signal intensity and the KK receiver. In addition, we propose a MIMO Volterra equalizer with crosstalk-dependent steps to mitigate XT and nonlinear degradation. Compared with the single-input single-output Volterra (SISO-Volterra) equalizer, the gains of received power and CSPR can be up to 0.7 dB and 1.3 dB respectively.

5.3.1 Experimental Setup and Analysis

The transmit signal of the KK system is a SSB optical signal, and i.e., the signal spectrum is on one side of the continuous wave (CW) tone. At the receiver, the KK algorithm reconstructs the full complex optical field information from the beating of signal with the CW tone by direct detection [51]. The reconstructing the optical field is based on the premise that it can satisfy the minimum phase condition [52], which means that the power of the CW tone is sufficiently high. We briefly review the essential algorithms of the KK receiver, and the details can be found in [42]. The optical field impinging upon the photodetector (PD) is expressed as

$$E(t) = E_s(t) + E_0(t)e^{i2\pi f_0 t}\left(f_0 \geq \frac{B}{2}\right) \tag{5.24}$$

where $E_s(t)$ represents the complex optical field of the signal, with the optical bandwidth of B. $E_s(t)$ representing a CW tone with amplitude $|E_0|$, which has a frequency approximating the center frequency of the signal. f_0 is the frequency difference between the center frequency of the signal and the CW tone after the frequency shift, which is called the frequency interval. The current I produced by the PD is

$$I(t) = |E(t)|^2 = |E_s(t)|^2 + 2\,Re\left(E_s(t)E_0^*(t)e^{i2\pi f_0 t}\right) + |E_0(t)|^2, \tag{5.25}$$

where the symbol (*) represents signal conjugation. Assuming that $|E_0|$ is large enough to ensure that the signal meets the minimum phase condition, the signal $E_s(t)$ can be reconstructed from the optical intensity $I(t)$ as

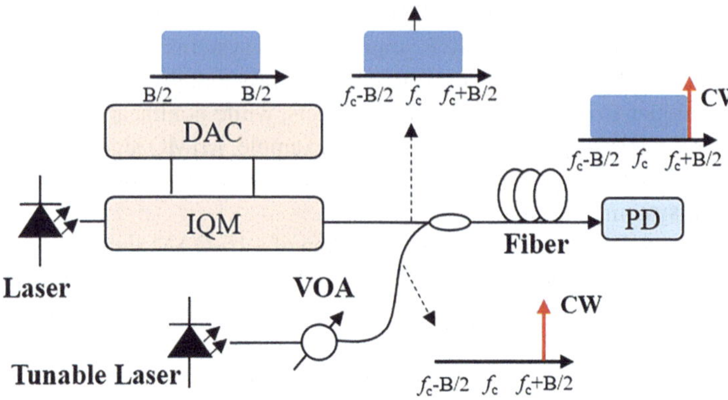

Fig. 5.13 Scheme of SSB optical signal generation

$$E_s(t) = \left\{ \sqrt{I(t)} e^{i\phi_e(t)} - E_0 \right\} e^{i2\pi f_o t} \tag{5.26}$$

$$\phi_e(t) = \frac{1}{2\pi} p.v. \int_{-\infty}^{\infty} \frac{log[I(t')]}{t - t'} dt' \tag{5.27}$$

where *p. v.* stands for Cauchy principal value.

In this section, the scheme shown in Fig. 5.13 is used to generate an SSB optical signal, and a tunable laser is employed to generate the CW tones for reconstructing the optical field information by the KK algorithm described above. Experimentally, it is able to alter and optimize the CSPR and frequency interval more easily by adjusting the tunable laser.

The experimental setup for the seven-core KK optical communication system shown in Fig. 5.14 CW is generated by the tunable laser source 1 (TLS1) with a linewidth of less than 100 kHz and output power less than 15 dBm. An I/Q modulator is then used to modulate the 10 GBaud 16-QAM signal with a 3 dB bandwidth of 33 GHz and an insertion loss of 6 dB. The signals are root-raised cosine (RRC) shaped with a roll-off of 0.2. An arbitrary waveform generator (AWG) of 64 GSa/s is used to convert two independent sequences of digital symbols into the electrical signals. An erbium-doped fiber amplifier (EDFA) with a noise figure of 5.5 dB and a variable optical attenuator (VOA) is used to adjust the signal launch power into a 1 × 2 coupler. The CW tone is generated by a TLS2 which can adjust the frequency. Another VOA adjusts the output power of carrier light, then coupled into the other side of the 1 × 2 coupler. The spectrum of the transmitted signal is shown in inset(i) of Fig. 5.14. We use a 1 × 8 coupler and the delay lines for decorrelation of different channels and then couple the seven-channel signals into the seven-core fiber through fan-in and fan-out (FIFO) modules. A fiber delay of about 20 ns is applied to core2, core4, and core7 respectively, and a delay of about 30 ns is applied to core1, core3, and core6. The inter-core skew with relation to the central core (core5) is measured to be −110 ps, −490 ps, 97 ps, 182 ps, −506 ps, and

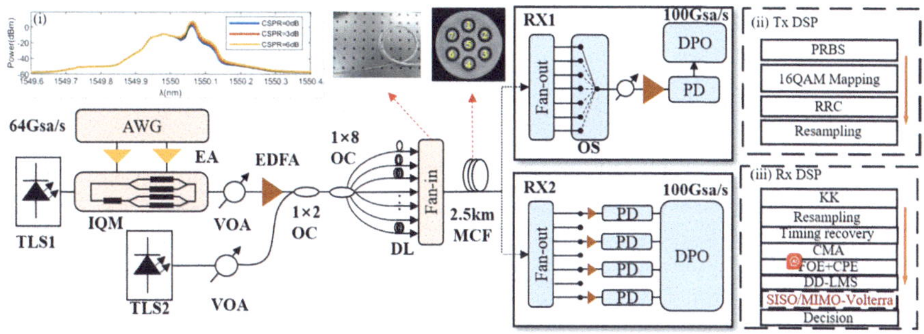

Fig. 5.14 Experimental setup for the seven-core KK 16-QAM coherent optical communication system (*TLS* tunable laser source, *AWG* arbitrary waveform generator, *IQM* I/Q modulator, *EA* electric amplifier, *OC* optical coupler, *VOA* variable optical attenuator, *DL* delay line, *EDFA* erbium-doped fiber amplifier, *OS* optical switch, *PD* photodetector, *DPO* digital phosphor oscilloscope). Inset (i) shows the spectrum of transmitted signals. Inset (ii, iii) show the transmitter and receiver DSP respectively

Table 5.1 Parameters and values of seven-core fiber and FIFO modules

	Parameters	Values
Trench-assisted homogeneous seven-core fiber	Loss @1550 nm(dB/km)	0.25
	Mode field diameter @1550 nm (μm)	9.5
	Core layer diameter (μm)	7.9
	Core spacing (μm)	41.5 ± 1.5
	Cladding diameter (μm)	150 ± 2
FIFO modules	Maximum insertion loss (dB)	1.5

133 ps (core1 → core7), and neighboring cores can be effectively decorrelated. Table 5.1 shows the parameters of seven-core fiber and FIFO modules.

The transmission link is composed of 2.5 km seven-core optical fiber. Figure 5.15 shows the distributed crosstalk and crosstalk matrix in a seven-core fiber and FIFO. Figure 5.15 a shows the temporal evolution of the crosstalk power with an average power of −45 dB generated by a CW light wave of 1550.13 nm within a 2.5 km 7-core MCF over 70 min, based on the method in [28]. The measured core pairs are core4 and core5. The XT fluctuations are reduced when amplified spontaneous emission (ASE) noise is being used. Figure 5.15 b shows the crosstalk matrix of the seven-core fiber with FIFO at a given time interval, using the continuous wave light emitted from one core, and the optical power of all cores is measured. The horizontal axis indicates the measured core label and the vertical axis denotes the input core label. The values in the cells of the figure are the decibels of energy transfer. The diagonal values of the matrix indicate the loss of each core. The maximum loss of core5 is 6.91 dB and the minimum loss of core7 is 4.37 dB.

At the receiver, we adopt both RX1 and RX2 reception schemes. In the RX1, the signals are demultiplexed and selected using a fan-out module and an optical switch

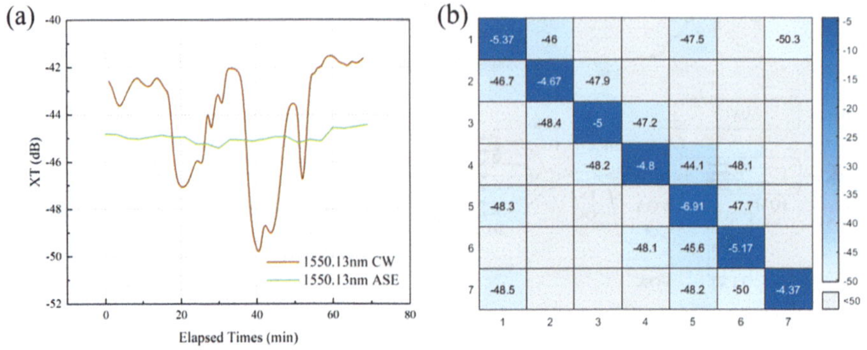

Fig. 5.15 Distributed crosstalk (**a**) and crosstalk matrix (**b**) in seven-core fiber and FIFO modules

(OS). Behind the OS, a VOA and an EDFA with fixed gain are employed to adjust the received power. A four-channel digital phosphor oscilloscope (DPO) with a sampling rate of 100 GSa/s, a bandwidth of 20 GHz, and a 50-GHz PD is used to receive the signal. In the RX2, four channels are selected to receive signal with four PDs, and the signals are fed to the DPO simultaneously for MIMO equalization. The offline DSP includes the KK algorithm, resampling, the constant modulus algorithm (CMA) equalization, frequency offset estimation (FOE), carrier phase estimation (CPE), the decision-directed least mean square (DD-LMS) algorithm, the SISO/MIMO-Volterra algorithm, and bit error rate (BER) calculation.

We first analyze the performance of the seven-core transmission system using the reception scheme RX1. Figure 5.16 presents the BER values at different signal powers for the KK receiver when the carrier optical power is 10 dBm, where the dashed line represents the result of optical transmission with only one core in the link, and i.e., there is no impact of XT; and the solid line represents the presence of XT. The performance gap between the dashed line and the solid line of the corresponding cores illustrates the impact of the above levels of crosstalk on signal performance. The transmission performance benefits from improved signal-to-noise ratio (SNR), with the signal power increasing from −8 dBm to 0 dBm. If the signal power increases to more than 0 dBm, fiber nonlinearity will be improved and the quality of the KK receiver will decrease, which will lead to the decrease in the BER performance. When the signal power is further increased to 6 dBm, the BERs of core5 and core3 are 8.51×10^{-3} and 1.44×10^{-3} respectively. In addition, the performance of the core varies in accordance with the diverse crosstalk levels and losses. It is noted that the performance gap between cores varies with signal power variations. It can be seen that from −6 dBm to 6 dBm, the performance difference between the cores for high signal power is greater than the that for low signal power. From Fig. 5.16, it can be predicted that for low power, the minimum power (core2) and maximum power (core5) for each core to reach below the FEC threshold are − 8 dBm and − 6 dBm respectively. In contrast, for the high power, the minimum power (core5) and maximum power (core7) for each core to return above the FEC threshold are 5 dBm and over 10 dBm respectively. Meanwhile, the performance

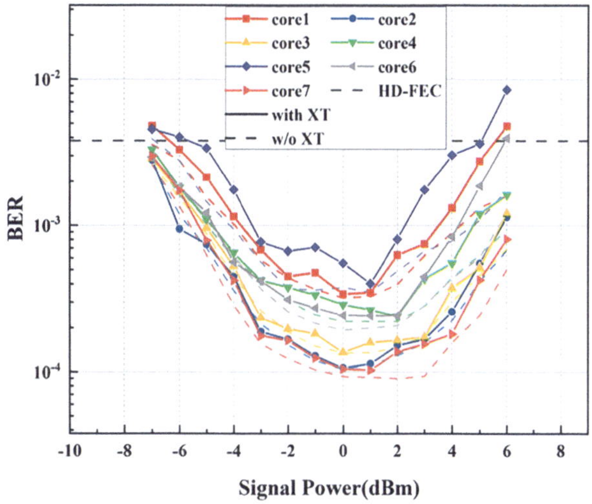

Fig. 5.16 BER versus signal power when carrier power is 10 dBm

difference between cores with XT at high signal power is greater than those without XT. The curves of all cores without XT tend to converge at high signal power, while the curves with XT are divergent. We predict that the crosstalk produced at low power is drowned out by the amplifier noise. The effect of XT is suppressed, whereas the impact of crosstalk at high power becomes significant. Furthermore, high signal power leads to stronger fiber nonlinearity, which interacts with crosstalk to compromise the performance of the SISO-Volterra equalizer, which makes the degradation of the BER greater than the effect of the XT. As the middle core, core 5 is affected by crosstalk between cores, which is consistent with the results in Fig. 5.16 and our conclusion as well.

Figure 5.17 describes the relationship between BER and CSPR when the signal power is 0 dBm. With the CSPR increasing from 3 dB to 13 dB, the BER of core7 decreases from 5.70×10^{-3} to 9.46×10^{-5} respectively. The maximum penalty of CSPR is 2.1 dB at the HD-FEC threshold (3.8×10^{-3}) because the crosstalk between core5 and core7 are different.

Figure 5.18 shows the relationship between BER and frequency interval when signal power is 0 dBm and CSPR is 10 dB. The bandwidth is 6 GHz when the RRC roll-off factor is 0.2. The signal power and the carrier power will be mixed, and the direct current (DC) component and the signal cannot be decoded when the frequency interval is <5 GHz. The BER decreases when the frequency interval increases from 5.625 GHz to 11.25 GHz. The BER values of core5 and core7 are 3.89×10^{-4} and 7.62×10^{-5} at a frequency interval of 11.25 GHz respectively. The BER increases from 7.62×10^{-5} to 6.69×10^{-3} due to the actual bandwidth limit of the oscilloscope when the frequency interval increases from 11.25 GHz to 13.75 GHz.

Figure 5.19 shows the relationship between BER and the received power when the signal power is 0 dBm and the CSPR is 10 dB. The BER will decrease from 8.87×10^{-2} to 6.68×10^{-5} when the received power is from 0 dBm to 7 dBm. The maximum penalty of received power is 1.2 dB at the HD-FEC threshold. The line

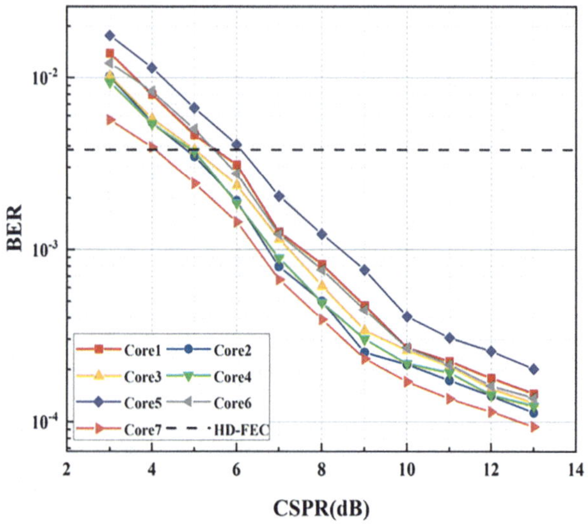

Fig. 5.17 BER as a function of CSPR when signal power is 0 dBm

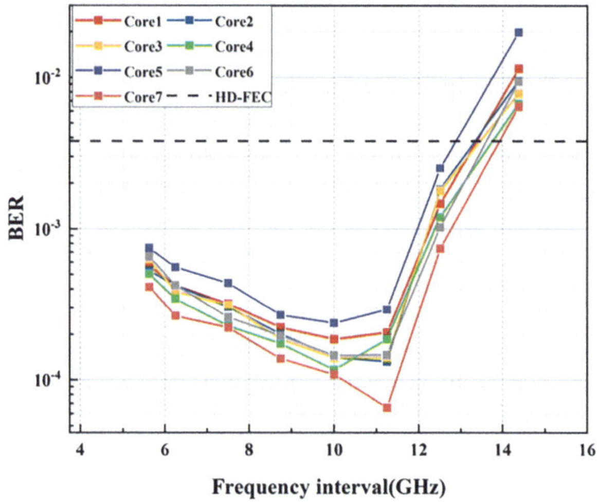

Fig. 5.18 BER as a function of frequency interval when carrier power is 10 dBm and the signal power is 0 dBm

rate of the system is 280 Gbits/s (7 × 10 GBaud × 4 bits/symbol). Taking into account the 7% FEC overhead, the system has a net bit rate of 261.7 Gbits/s.

5.3.2 XT-MIMO-Volterra Nonlinear Equalization Algorithm

A MIMO-Volterra equalization algorithm is proposed for compensating the nonlinear impairments, the power impairments, and XT in multicore fiber transmission systems. In multicore fibers, the coupling effects between modes in adjacent

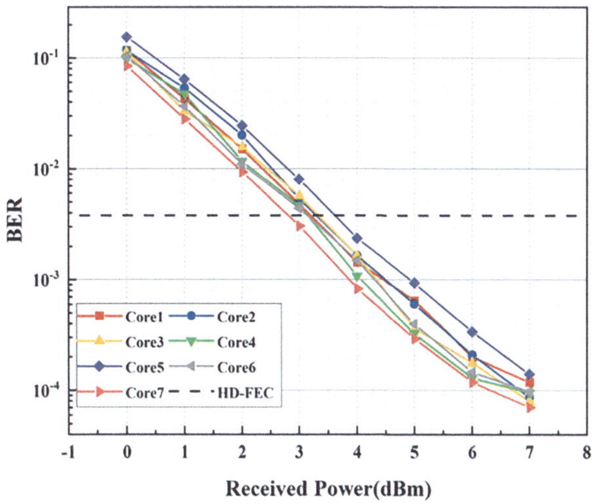

Fig. 5.19 BER as a function of frequency interval when carrier power is 10 dBm and the signal power is 0 dBm

cores lead to XT, which limit the capacity and transmission distance of the system. The input and output of the multicore system are a complex nonlinear relationship due to the combined effects of the XT in the multicore optical fiber and the nonlinear impairment of the devices [53–55]. To solve the problem of the nonlinear effects, the Volterra series can extract the nonlinear impairment characteristics of the device and the probability model of the nonlinear system. The Volterra series provides a complete description of the nonlinear impairments caused by devices such as electrical amplifiers, I/Q modulators, and PDs in the system.

The input–output relationship of the third-order limited-band Volterra model for the $_{i\text{-th}}$ fiber core can be expressed as

$$y_i^{'}(n) = \sum_{p=0}^{L_1} h^{(1)}(p)x_i(n-p) + \sum_{p=0}^{L_2}\sum_{q=p}^{L_2} h^{(2)}(p,q)x_i(n-p)x_i(n-q)$$
$$+\sum_{p=0}^{L_3}\sum_{q=p}^{L_3} h^{(3)}(p,q,l)x_i(n-p)x_i(n-q)x_i^*(n-l). \tag{5.28}$$

where $x_i(n)$ and $y_i'(n)$ are the input and output symbols of the $i-th$ core respectively. n is the index of signals. L_j is the memory length of the $j-th$ order. $h^{(j)}$ is the $j-th$ kernel of the Volterra filter. The first term of Eq. (5.28) represents a linear filter, while the other terms represent nonlinear terms. In the fiber optic transmission system, only the first and third terms are considered due to the fact that the second-order nonlinear memory length is 0. Then, the XT and nonlinearity of the MCF system can be represented by a nonlinear MIMO channel based on the Volterra series, and the input–output relationship of the MCF system can be expressed as

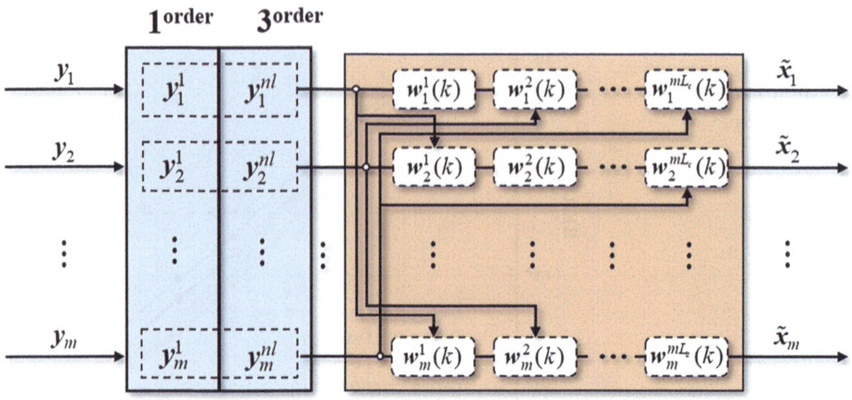

Fig. 5.20 Block diagram of the first-order and third-order MIMO-Volterra equalizers

$$\begin{bmatrix} y_1(k) \\ y_2(k) \\ \cdots \\ y_m(k) \end{bmatrix} = \begin{bmatrix} C_{11}(k) & C_{21}(k) & \cdots & C_{m1}(k) \\ C_{12}(k) & C_{22}(k) & & C_{m2}(k) \\ \cdots & & \cdots & \\ C_{1m}(k) & C_{2m}(k) & & C_{mm}(k) \end{bmatrix} \begin{bmatrix} y_1'(k) \\ y_2'(k) \\ \cdots \\ y_m'(k) \end{bmatrix}. \tag{5.29}$$

where $y_i(k)$ denotes the k-th output symbol of the i-th core of the MCF system; $C_{ij}(k)$ $(i, j = 1, 2, \ldots, m)$ is the signal crosstalk coefficient of k-th time interval. m represents the index of cores.

In this section, the nonlinear MIMO equalization algorithm based on the Volterra series and the characteristic coefficient of XT is proposed to compensate for linear crosstalk and nonlinear impairments in space division multiplexed systems. The MIMO-Volterra equalizer consists of linear equalization branches and nonlinear equalization branches, which are completed by the first-order and third-order terms of the Volterra equalizer. The block diagram of the equalizer is shown in Fig. 5.20. The MIMO nonlinear equalization scheme based on the Volterra series improves the noise tolerance of multicore fiber transmission systems. Figure 5.21 shows the diagram of the XT-MIMO-Volterra nonlinear equalization algorithm. The specific algorithm flow is as follows.

(1) Assume that the 16-QAM signal data matrix for all cores input to the equalizer is represented as

$$D = \begin{bmatrix} y_1(1) & y_1(2) & \cdots & y_1(n) \\ y_2(1) & y_2(2) & \cdots & y_2(n) \\ \cdots & \cdots & \cdots & \cdots \\ y_m(1) & y_m(2) & \cdots & y_m(n) \end{bmatrix} = \begin{bmatrix} y_1 \\ y_2 \\ \cdots \\ y_m \end{bmatrix} \tag{5.30}$$

where $y_m(n)$ represents signal and each channel includes the training signal and the payload signal.

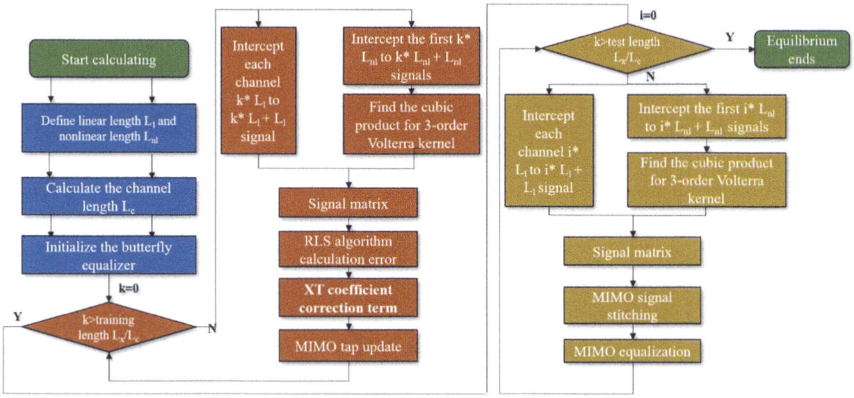

Fig. 5.21 Flow diagram of the XT-MIMO nonlinear equalization algorithm based on the Volterra series

(2) Calculate the third-order nonlinear length as

$$L_{nl} = \sum_{v_3=v_1}^{L_3} \sum_{v_2=v_3}^{L_3} \sum_{v_1=1}^{L_3} = \frac{1}{12}[3L_3(L_3+1) + L_3(L_3+1)(2L_3+1)]. \quad (5.31)$$

(3) Calculate the total channel length as

$$L_c = L_1 + L_{nl}. \quad (5.32)$$

(4) Initialize the butterfly equalizer, and the tap coefficient of the equalizer can be expressed as

$$w(k) = \begin{bmatrix} w_1^0(k) & w_2^0(k) & \dots & w_m^0(k) \\ w_1^1(k) & w_2^1(k) & \dots & w_m^1(k) \\ \dots & \dots & \dots & \dots \\ w_1^{mL_c}(k) & w_2^{mL_c}(k) & \dots & w_m^{mL_c}(k) \end{bmatrix}$$
$$= \begin{bmatrix} w_{11}(k) & w_{12}(k) & \dots & w_{1m}(k) \\ w_{21}(k) & w_{22}(k) & \dots & w_{2m}(k) \\ \dots & \dots & \dots & \dots \\ w_{m1}(k) & w_{m2}(k) & \dots & w_{mm}(k) \end{bmatrix} \quad (5.33)$$

Let $w_i^{[H_1]}(0)(i=1,2,\ldots,m)$ is 1, where $H_1 = \frac{L_1}{2}$ and the other terms of $w(0)$ are 0. $w_{ij}(k)(i,j=1,2,\ldots,m)$ is a column vector of length L_c.

(5) Intercept the $(k + 1)[H_1] - [H_1]$ to $(k + 1)[H_1] + [H_1]$ signals of each core from $k = 1$ and arrange them in reverse order to obtain the linear input sequence of the equalizer for each channel as

$$y_i^1(k) = \begin{bmatrix} y_i((k+1)\lceil H_1 \rceil + \lfloor H_1 \rfloor) \\ \cdots \\ y_i((k+1)\lceil H_1 \rceil - \lfloor H_1 \rfloor) \end{bmatrix}. \tag{5.34}$$

Intercept the signals from $(k + 1)[H_1] - [H_3]$ to $(k + 1)[H_3] + [H_3]$ of each core and arrange them in reverse order, using Eq. (5.35) to find the cubic product of the input signals corresponding to the third-order Volterra kernel for each channel.

$$y_i^{nl}(k) = \begin{bmatrix} y_i(A)y_i(A)y_i(A) \\ \cdots \\ y_i(A - v_3)y_i(A - v_2)y_i(A - v_1) \\ \cdots \\ y_i(B)y_i(B)y_i(B) \end{bmatrix}, \tag{5.35}$$

Where

$$A = (k+1)\lceil H_3 \rceil + \lfloor H_3 \rfloor$$
$$B = (k+1)\lceil H_3 \rceil - \lfloor H_3 \rfloor.$$

(6) Splice the linear sequence with a third order to obtain the input sequence for each channel, shown as

$$y_i(k) = \begin{bmatrix} y_i^1(k) \\ y_i^{nl}(k) \end{bmatrix}. \tag{5.36}$$

(7) Splice the MIMO equalizer input sequence of each channel into a MIMO input sequence, as shown in Eq. (5.37).

$$y(k) = \begin{bmatrix} y_1(k) & y_2(k) & \cdots & y_m(k) \\ y_2(k) & y_1(k) & \cdots & y_1(k) \\ \cdots & \cdots & \cdots & \cdots \\ y_m(k) & y_m(k) & \cdots & y_{m-1}(k) \end{bmatrix}. \tag{5.37}$$

(8) Calculate the MIMO output signal as

$$\widetilde{x}(k) = w^T(k-1)y(k) = \begin{bmatrix} \widetilde{x}_1(k) \\ \widetilde{x}_2(k) \\ \cdots \\ \widetilde{x}_m(k) \end{bmatrix}. \tag{5.38}$$

(9) Calculate the error as

$$\varepsilon(k) = d(k) - \widetilde{x}(k) = \begin{bmatrix} \varepsilon_1(k) \\ \varepsilon_2(k) \\ \cdots \\ \varepsilon_m(k) \end{bmatrix}. \tag{5.39}$$

(10) Update the inverse of the correlation matrix as [42]:

$$\mathbf{R}_{D,i}^{-1}(k) = \frac{1}{\lambda}\left[\mathbf{R}_{D,i}^{-1}(k-1) - \frac{\mathbf{R}_{D,i}^{-1}(k-1)\mathbf{y}_i(k)\mathbf{y}_i^T(k)\mathbf{R}_{D,i}^{-1}(k-1)}{\lambda + \mathbf{y}_i^T(k)\mathbf{R}_{D,i}^{-1}(k-1)\mathbf{y}_i(k)}\right]. \tag{5.40}$$

where $R_{D,i}^{-1}(k)(i=1,2,\ldots,m)$ is the inverse of the correlation matrix, and $R_{D,i}^{-1}(k)$ is initialized to a unit matrix. $0 < \lambda \leq 1$ is called the forgetting factor, which has some forgetting effect on historical data. In this section, $\lambda = 1$, which means no forgetting effect exists.

Calculate the crosstalk coefficient matrix according to the measured crosstalk matrix as

$$\mathbf{XT}_{opt}(k) = \begin{bmatrix} XT_{opt.11}(k) & XT_{opt.12}(k) & \ldots & XT_{opt.1m}(k) \\ XT_{opt.21}(k) & XT_{opt.22}(k) & \ldots & XT_{opt.2m}(k) \\ \cdots & \cdots & \cdots & \cdots \\ XT_{opt.m1}(k) & XT_{opt.m2}(k) & \ldots & XT_{opt.mm}(k) \end{bmatrix}, \tag{5.41}$$

Where

$$XT_{opt.ij}(k) = \rho \frac{XT_{ij}(k) - XT_{min}(k)}{XT_{max}(k) - XT_{min}(k)} (i,j = 1,2,\ldots,m). \tag{5.42}$$

$XT_{ij}(k)$ represents the value of the measured crosstalk level of k-th time interval, and $XT_{max}(k)$ and $XT_{min}(k)$ are the maximum and minimum values in the crosstalk matrix respectively. ρ is the scaling factor. For example, in Fig. 5.15 b, $XT_{12}(k) = 10^{-46.7/10}$, $XT_{max}(k) = 10^{-4.67/10}$ and $XT_{min}(k) = 10^{-50.3/10}$, which is negligible for less than -50 dB (gray cells in the crosstalk matrix), corresponding to $XT_{opt,\ ij}(k)$ of 0. Since the inter-core XT changes relatively slowly, the crosstalk matrix is then collected once in one minute. We use the average crosstalk power in one minute for the calculation of the algorithm crosstalk coefficient, i.e., $XT_{ij}(1) = XT_{ij}(2) = \ldots = XT_{ij}(k)$. The transmitted signal is processed with the crosstalk matrix in the corresponding time interval. Considering that $XT_{opt}(k)$ is a singular matrix, the inverse matrix of $XT_{opt}(k)$ is

$$\mathbf{XT}_{opt}^{-1}(k) = \begin{bmatrix} XT'_{opt.11}(k) & XT'_{opt.12}(k) & \ldots & XT'_{opt.1m}(k) \\ XT'_{opt.21}(k) & XT'_{opt.22}(k) & \ldots & XT'_{opt.2m}(k) \\ \ldots & \ldots & \ldots & \ldots \\ XT'_{opt.m1}(k) & XT'_{opt.m2}(k) & \ldots & XT'_{opt.mm}(k) \end{bmatrix}, \tag{5.43}$$

(11) Update the tap coefficient of the equalizer using Eqs. (5.43) and (5.44):

$$\boldsymbol{w}_{ij}(k) = \boldsymbol{w}_{ij}(k-1) + XT'_{opt.ij}\frac{\boldsymbol{R}_{D,i}^{-1}(k-1)\boldsymbol{y}_i(k)}{\lambda + \boldsymbol{y}_i^T(k)\boldsymbol{R}_{D,i}^{-1}(k-1)\boldsymbol{y}_i(k)}\epsilon_i(k). \tag{5.44}$$

See Appendix A for a detailed explanation of the derivation process.

(12) Repeat steps 5–12 until the end of the training signal is updated and converged.

(13) Extract the payload signal and repeat steps 5–8 until the end of the payload signal is updated, and complete the MIMO equalization.

5.3.3 Experimental Results and Performance Analysis of Algorithm

Figure 5.22 a shows the constellation diagram of the signal. The signal mainly suffers from the nonlinear impairment of the device including the FIFO modules and the influence of crosstalk in the transmission link. The first- and third-order memory lengths of Volterra filters are optimized to 21 and 5 respectively. The performance of the signal improved after adopting the SISO-Volterra algorithm, MIMO-Volterra algorithm, and XT-MIMO-Volterra algorithm respectively. Figure 5.22 b shows the mean square error (MSE) curve of the signal. The peaks of the MSE after passing through the SISO-Volterra algorithm, the MIMO-Volterra algorithm, and the XT-MIMO-Volterra algorithm are 6, 2.7, and 1.6 respectively. It can be seen that the MIMO-Volterra algorithm can effectively restrain the signal impairments, and the XT-MIMO-Volterra algorithm shows the best performance for equalizing the signal impairments compared with two other algorithms.

Figure 5.23 shows the experimental results for the seven-core transmission system with the KK receiver. Linear coordinates are employed for the y-axis in Fig. 5.23 a, and logarithmic coordinates are used in Fig. 5.23 b–d to enhance readability. Figure 5.23 a shows the relationship between BER and signal power

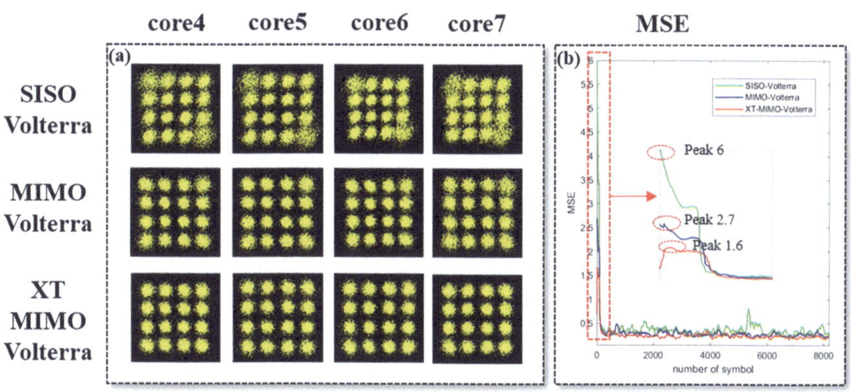

Fig. 5.22 Experimental results: (**a**) the constellation diagram and (**b**) the mean square error curve of the signal after passing through the SISO-Volterra algorithm, the MIMO-Volterra algorithm, and the XT-MIMO-Volterra algorithm, respectively

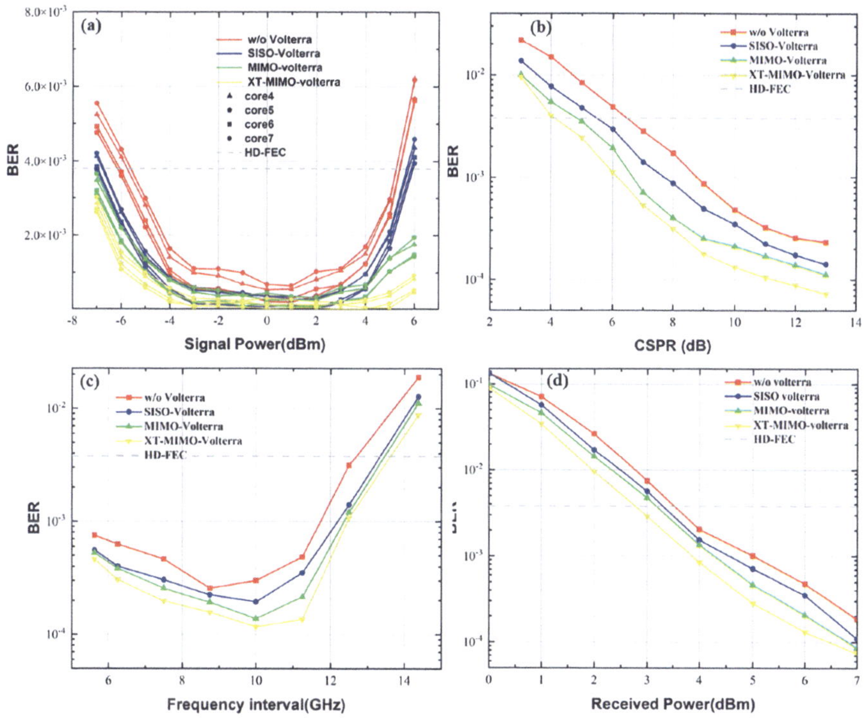

Fig. 5.23 Experimental results. (**a**) BER as a function of signal power when the carrier power is 10 dBm, (**b**) BER as a function of CSPR when the carrier power is 10 dBm, (**c**) the relationship between BER with frequency interval (**c**) and received power (**d**) respectively

when the carrier optical power is 10 dBm. The observed signal power ranges from −7 dBm to 6 dBm, and the BER decreases with the increase of signal power ranging from −7 dBm to 0 dBm. As the signal power increases from 0 dBm to 6 dBm, the BER performance is getting worse, which is mainly affected by the CSPR and the quality of the KK receiver. For low CSPR, BER performance is more immensely improved by the MIMO-Volterra equalizer. For instance, the BER of MIMO-Volterra and XT-MIMO-Volterra are both below the FEC threshold when the signal power increases beyond 0 dBm, while w/o Volterra and SISO-Volterra are above the FEC threshold at 5.2 dBm and 5.7 dBm respectively. There is little correlation between the difference in performance of cores and algorithms. Therefore, the average BER of the four channels (core4, core5, core6, and core7) is used to clearly compare the performances between different algorithms in Fig. 5.23 b–d.

The signal power is set at 0 dBm in Fig. 5.23 b. The measured CSPR varies from 3 dB to 13 dB. When the CSPR is >13 dB, the optical power exceeds the normal operating range of the seven-core fan-in module. As the CSPR rises, the XT-MIMO-Volterra algorithm will show the optimal performance. When the HD-FEC threshold is achieved, the CSPRs of w/o Volterra, SISO-Volterra, MIMO-Volterra, and XT-MIMO-Volterra are 6.5 dB, 5.4 dB, 4.8 dB, and 4.1 dB, respectively. Compared with MIMO-Volterra and SISO-Volterra, the XT-MIMO-Volterra algorithm can reduce the CSPR by 0.7 dB and 1.3 dB respectively.

In Fig. 5.23 c, the carrier optical power is fixed to 10 dBm and the observed frequency interval range is from 5.625 GHz to 14.375 GHz. The 10 GBaud signal and carrier are aliased and cannot be demodulated typically when the frequency interval is less than 5 GHz. The performance decreases with the increase in the frequency interval from 5.625 GHz to 11.25 GHz. For frequency interval = 11.25 GHz, the BERs of w/o Volterra, SISO-Volterra, MIMO-Volterra, and XT-MIMO-Volterra are 4.88×10^{-4}, 3.55×10^{-4}, 2.14×10^{-4}, and 1.36×10^{-4} respectively. As the frequency spacing increases from 11.25 GHz to 14.375 GHz, the BER performance degrades with the frequency spacing due to the 3-dB bandwidth of the oscilloscope reaching 20 GHz, which limits the performance.

Figure 5.23 d shows the relationship between BER and received power for a signal power of 0 dBm and a carrier optical power of 10 dBm. The received power in the seven-core experimental system with KK receiver is from 0 dBm to 7 dBm, and the BER decreases with the increase in CSPR. For the received power = 7 dBm, the BERs of w/o Volterra, SISO-Volterra, MIMO-Volterra, and XT-MIMO-Volterra are 1.83×10^{-4}, 1.09×10^{-4}, 9.15×10^{-5} and 7.62×10^{-5} respectively. The HD-FEC threshold can be achieved when the received power is 3.5 dBm to 7 dBm. At the HD-FEC threshold, the gain of the XT-MIMO-Volterra algorithm improved by 0.5 dB and 0.7 dB compared with the MIMO-Volterra and SISO-Volterra algorithms respectively.

In this section, the experiment of a 261.7 Gbit/s 16-QAM signal transmitting a 2.5 km seven-core fiber with a KK receiver was demonstrated. We investigated the transmission characteristics of CSPR, the signal power, and the protection interval on seven-core fiber. The maximum penalty of CSPR is 2.1 dB at the HD-FEC threshold. The core transmission performance shows primarily a dependence on crosstalk and losses of different cores. The performance gap between the cores shows a differentiation with the change of signal power. We predict that this is related to the resulting power of XT, the nonlinearity and the amplifier noise. Moreover, XT-MIMO-Volterra is proposed to reduce inter-core crosstalk and nonlinear degradation. Compared with the MIMO-Volterra and SISO-Volterra algorithms, CSPR can be reduced either by 0.7 dB or by 1.3 dB respectively, and the received power gain of XT-MIMO-Volterra algorithm is improved either by 0.5 dB or 0.7 dB respectively. In this section, the crosstalk level of the short-reach MCF link is relatively low, and the XT-MIMO-Volterra equalizer can be verified in the long-haul link in the future.

5.4 MCF-SHCD System Based on PS MD Modulation

With emerging Internet services like cloud computing, online education, and real-time HD video, there is an explosive growth in short-reach data traffic transmission. Considering the strict system requirements for power consumption and module cost, intensity modulated direct detection (IM-DD) has been the most popular approach for short-reach optical transmission. As the data rate exceeds Tbit/s, the IM-DD system will encounter a bottleneck due to limited bandwidth, limited spectral efficiency (SE) and low dispersion (CD) tolerance [43]. However, full coherent systems are not straightforward to employ for short-reach due to the expensive cost and high-power consumption. Therefore, some research on SHCD becomes popular.

The self-homodyne coherent detection (SHCD) system is a promising approach for achieving 1.6 Tbit/s and higher transmission rates in future short-reach optical links [44–46]. In SHCD systems, the unmodulated carrier is transmitted as a pilot tone (PT) together with the coherent signal from the transmitter, and the PT is available for a local oscillator at the receiver, eliminating the requirement for carrier frequency synchronization between the transmitter and the receiver. The transceiver laser power consumption and requirements for digital signal processing are minimized, as a result, the high capacity and substantial power efficiency are achieved. In SHCD systems, there is an additional channel dimension that is required for the transmission of PTs, which results in a reduction of spectral efficiency (SE). For instance, if the transmission of a PT is based on one polarization of the light wave, or the addition of an extra fiber, the SE will reduce by 50%. The widely investigated space division multiplexing (SDM) technology [47] moderately alleviates this issue. SDM fibers with multiple cores and various spatial modes increase the number and dimension of transmission channels in a single fiber [48, 49]. For the SDM extension of N spatial channels, the SE is reduced by 1\N with the transmission of PTs using one core or one spatial mode.

On the other hand, once the SE and flexibility are close to the Shannon limit in SHCD systems, probabilistic shaping (PS)-based ultra-high-order modulation (UHM) techniques can be implemented. It has been demonstrated to perform many experiments on UHM signal [50–60]. Nevertheless, with the increasing of the modulation format order, the minimum Euclidean distance (MED) is reduced and the noise immunity decreases. Therefore, it is crucial to propose a new modulation format to extend the MED of constellation points to improve the system performance. The spectrally efficient multidimensional (MD) modulation format provides higher coding gains and greater demodulation margins [50]. Four-dimensional (4D) modulation formats have been extensively researched based on two orthogonal phase components and two polarizations of the optical field [61, 62]. A relatively well-known approach is the coded modulation based on the Ungerboeck's set-partitioning (SP) scheme, which increases the MED between symbols by XOR operations to produce parity bits [26, 28, 63]. However, it is not directly cascaded able with PS due to the presence of inter-symbol coding operations. Several recent MD signaling schemes for PS [64–66] have been reported to provide higher tolerance towards fiber nonlinearities, but there is still a requirement for flexible platforms to provide rate adaptation of MD signals based on SP. Therefore, we proposed a rate-adaptive four-dimensional (4D) PS UHM modulation format based on amplitude translation (AT) in a previous work [67].

In this section, we employ MD modulation formats for the MCF-SHCD system. Specifically, 4D to 12D coded modulation schemes are constructed at the transmitter based on the amplitude translation (AT)-SP principle and PAS structure, and the redundancy of the MD signals is split and transmitted in a core with the PT. At the receiver, the PT and redundancy signals are extracted for SHCD and decoding of the parity, respectively. Hard and soft decoders for MD signals are introduced, and the multidimensional decoder can be directly cascaded with the PAS decoder. Furthermore, we implement 20 GBaud 4D/8D/12D PS-16384-QAM transmission over 22.5 km uncoupled MCF to analyze the performance of the proposed scheme. The experimental results show that the SE gains of 0.84 bit/s/Hz, 1.49 bit/s/Hz and 2 bit/s/Hz can be obtained for hard decision 4D/8D/12D PS-16384-QAM, respectively, and the SE gains of 1.7 bit/s/Hz, 2.62 bit/s/Hz and 3.31 bit/s/Hz can be obtained for soft decision 4D/8D/12D PS-16384-QAM, compared with the PDM signal over 22.5 km transmission. The maximum SE of the MD-MCF-SHCD system with the soft decision is 52.59 bit/s/Hz (17.5 bit/s/Hz/4D-sym), and a bit rate of 1.06 Tbit/s (354.1 Gbit/s/core).

5.4.1 System Model

We obtain SHCD systems based on multidimensional modulation formats, and the system structure is shown in Fig. 5.24. At the transmitter, uniformly distributed information bits generated by a discrete binary memoryless source (DBMS) are fed into the probabilistic amplitude shaping (PAS) for each of the N paths, which produces symbols following the Maxwell-Boltzmann (MB) distribution. In a one-path PAS, the input bit sequence is divided into a bit sequence b^l of length 1 and a bit sequence b^m of length m. The sequence b^l is fed into the distribution

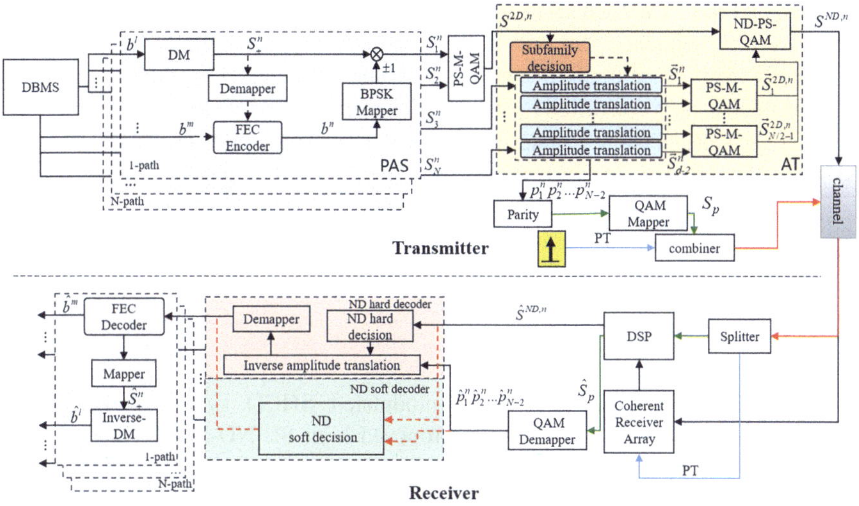

Fig. 5.24 Structure of SHCD system based on ND modulation

matcher (DM) to produce non-uniformly distributed positive amplitude symbols S_+^n, with the alphabet A = {1, 3, . . ., √M}. The symbols S_+^n from the DM are then reflected in a demapper as sequence according to the Gray mapping rule, which are combined with a sequence of bits b^m to be fed into the forward error correction (FEC) encoder to produce parity. The sequence of parity produced by the FEC encoder together with the sequence b^m forms a new sequence b^n. Then b^n is converted by the binary phase shift keying (BPSK) mapper from bit to n amplitudes, i.e., 0 is mapped to −1 and 1 is mapped to 1. These ±1 act as signs for S_+^n to form √M-PAM signals S^n. After PAS, sequence of non-uniformly distributed symbols sequences S_1^n, S_2^n, . . ., S_N^n of N paths is formed. A sequence of 2D PS-M-QAM symbols $S^{2D,\,n}$ is comprised of the symbol sequences S_1^n and S_2^n for path 1 and path 2. $S^{2D,\,n}$ is the basis for SFC determination as the 2D base symbol sequence of the ND symbol sequence. The current SFC matched to a ND symbol is determined according to the 2D subset in which $S^{2D,\,n}$ is located, and the symbol of the corresponding path is amplitude translated. The translated symbol sequences of third to N-th dimensions are paired to form 2D PS-M-QAM symbol sequences $\vec{S}_1^{2D,n}$, $\vec{S}_2^{2D,n}$, . . ., $\vec{S}_N^{2D,n}$. Eventually, the signals of all dimensions are combined to form an effective ND modulated signal $S^{ND,\,n}$. On the other hand, N − 2 parity bits p_1^n, p_2^n, . . ., p_{N-2}^n are produced for AT of each ND symbol, and these parity bits are mapped to additional uniform QAM symbol S_p. This QAM signal and the optical PT are combined to form a mixed signal with no interference, which is transmitted with the $S^{ND,\,n}$ over the channel.

At the receiver, the transmitted ND symbols are received via a homodyne coherent receiver array, and the mixed signal is split into two branches via a splitter. The uniform QAM signal branch carrying ND parity is received by another coherent receiver, and the PT branch serves as the LO for all coherent receivers. The received QAM symbol sequence ˆSp and the ND symbol sequence ˆSND, n are ideally eliminated from other impairments by DSP, since the signal is only impaired by Gaussian noise. $\widehat{S}_p$ is converted by the QAM demapper to parity bits $\widehat{p}_1^n, \widehat{p}_2^n, \ldots, \widehat{p}_{N-2}^n$ for HDD or soft-decision decoder (SDD) of the recovered ND signal. The decided ND signal is divided into N paths for FEC decoding, mapping and Inverse-DM, respectively, to finally obtain the transmitted binary bits $\widehat{b}^l$ and $\widehat{b}^m$.

5.4.2 Experimental Setup

We demonstrated an ND-UHD-QAM-modulated SHCD system based on an uncoupled 7-core fiber to validate the proposed high SE ND-SHCD transmission technique. The experimental setup of the SHCD experimental system is shown in Fig. 5.25. The light source is a single narrow linewidth laser with a linewidth of 1 kHz operating at 1550 nm. Continuous wave light is emitted with a power of 17.9 dBm and split into three paths via a coupler. The light in the first path is fed to an erbium-doped fiber amplifier (EDFA) with a noise figure of 5.5 dB, amplifying the optical power to 20 dBm. The amplified light is modulated by a dual polarization integrated coherent drive modulator (CDM). The modulator is driven by a four-channel arbitrary waveform generator (AWG) with a nominal resolution of 8 bits and a 3-dB bandwidth of 40 GHz. The AWG produces 20 GBaud ND PS-16384-QAM, operating at 120 GSa/s. The ND PS-16384-QAM employs a Nyquist root-raised cosine (RRC) filter with a roll-off coefficient of 0.01, resulting in a signal bandwidth of 20.2 GHz. Then, the light is pre-amplified by an EDFA, and the amplified light is divided into three paths by a coupler. The optical signals of each path are fed to the fan-in/fan-out device (FIFO) of the MCF and transmitted after a delay of different lengths. The ND signal is carried by the in-phase and quadrature components of two orthogonal polarizations of 1-3 fiber cores. For example, the 8D signal consists of the components of 4 dimensions in core 1 and core 2. Since there is only one four-channel AWG, the components of the ND signal in each core are sent for different periods of time. The second path from the laser is modulated by an AWG-driven IQ modulator, which produces the corresponding parity signal, mapped by the parity

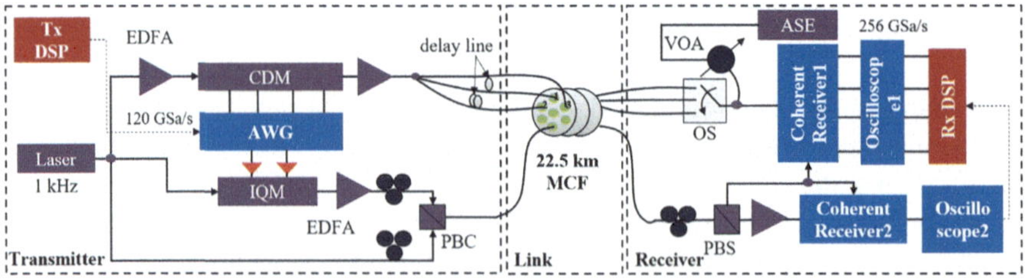

Fig. 5.25 The experiment setup of ND-MCF-SHCD transmission

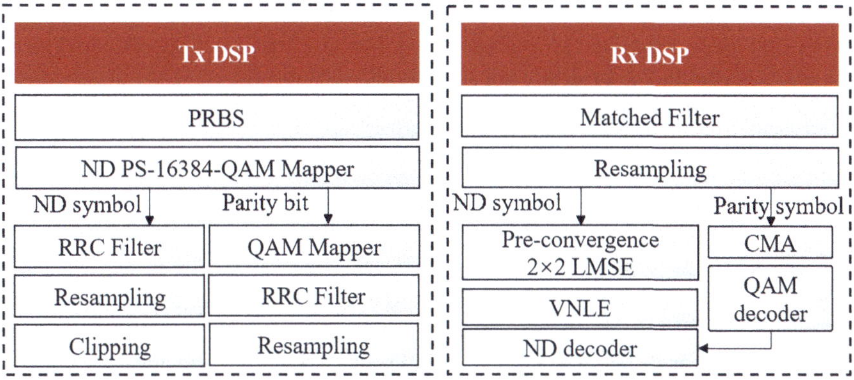

Fig. 5.26 The DSP of ND-MCF-SHCD transmission

bits of the ND modulation. The parity signal is amplified by an EDFA and rotated in the polarization by a polarization controller (PC). The third path from the laser is considered as a PT and is combined with the modulated light of the second path by a polarization beam combiner (PBC) after a PC and sent to another core of the MCF for transmission. The PCs are manually adjusted at fixed time intervals to achieve polarization alignment of the pilot and parity signals for optimal receiver performance. The transmission fiber is a single 22.5-km span of 7-core fiber, with a loss of 0.25 dB/km and inter-core crosstalk of less than 55 dB.

At the receiver, the signal light from the three cores is selected by an optical switch (OS) that loads with noise by an amplified spontaneous emission (ASE) noise source and a variable optical attenuator (VOA). It is subsequently received by a dual polarization self-homodyne coherent receiver and an oscilloscope. The oscilloscope operates at 256 GSa/s and the waveform data is acquired for the Rx DSP. The PT/parity hybrid signal light is adjusted by a PC to rotate the polarization and is split by a polarizing beam splitter (PBS). The parity branch is received by another coherent receiver and performs DSP with the ND signal. And the PT branch is used for the LO of the two coherent receivers. Figure 5.26 shows the DSP of the transmitter and receiver. In the Tx DSP, the pseudo-random binary sequence (PRBS) is mapped by the ND mapper to the ND PS-16384-QAM signal with its corresponding parity. The ND symbols are RRC filtered, resampled and clipped to produce a 20 GBaud signal which is loaded into the AWG. Parity bits are mapped to 20 GBaud QPSK/64QAM according to 4D/8D respectively. For 12D, it is unacceptable for noise tolerance at 20 GBaud due to mapping parity bits to 1024-QAM. Therefore, the parity bits of 12D are transmitted with 40 GBaud 32-QAM. The parity signal is RRC-filtered, resampled and loaded into the AWG. In the Rx DSP, matched filtering and resampling are performed for the received ND symbols and parity signals. A 2× oversampling pre-convergence blind least mean square equalizer (LMSE) is utilized to perform polarization demultiplexing and compensate for inter-symbol interference (ISI). A third-order Volterra nonlinear equalizer (VNLE) is employed, where the first-order, second-order, and third-order memory lengths are

201, 7 and 7 respectively. The SHCD and the narrow linewidth of 1 kHz suppress carrier frequency offset and phase noise, and thus no carrier recovery is performed. The parity signal flows through a constant modulus algorithm (CMA) equalizer and decoder and is finally used to decode the ND signal.

5.4.3 Experiment Results

Figure 5.27 a–c show the NGMI results for 4D/8D/12D signal over 22.5 km MCF. The results in Fig. 5.9 a are evaluated in one core, where the PDM (2D) signals with HDD and SDD for Rc allows maximum HNDs of 17.4 bit/4D-sym and 17.9 bit/4D-sym. While the maximum H_{ND} of 4D signals with HDD and SDD is extended to 17.7 bit/4D-sym and 19.1 bit/4D-sym respectively, which implies higher SE. The signals are evaluated for two fiber cores in Fig. 5.19 b. For the 2D signals with HDD and SDD, the maximum H_{ND} can be 35.7 bit/8D-sym and 37.4 bit/8D-sym, while the maximum H_{ND} for the 8D signals with HDD and SDD is extended to 36.8 bit/8D-sym and 40.3 bit/8D-sym. The signals are evaluated for three fiber cores in Fig. 5.19 c. The maximum H_{ND} allowed for the PDM signals with HDD and SDD in Fig. 5.19 c are 53.4 bit/12D-sym and 56.9 bit/12D-sym. And the maximum HND with HDD and SDD is extended to 56.7 bit/12D-sym and 60.9 bit/12D-sym, respectively. Meanwhile, the gradient colors represent the NGMI results for each core and each polarization component. It is observed that the NGMI of each component of the PDM signal is more divergent, while the components of the ND signal substantially converge, which implies that there is an averaging effect of the ND signal for each component. Figure 5.27 d–f illustrate the AIR for 4D/8D/12D signaling over

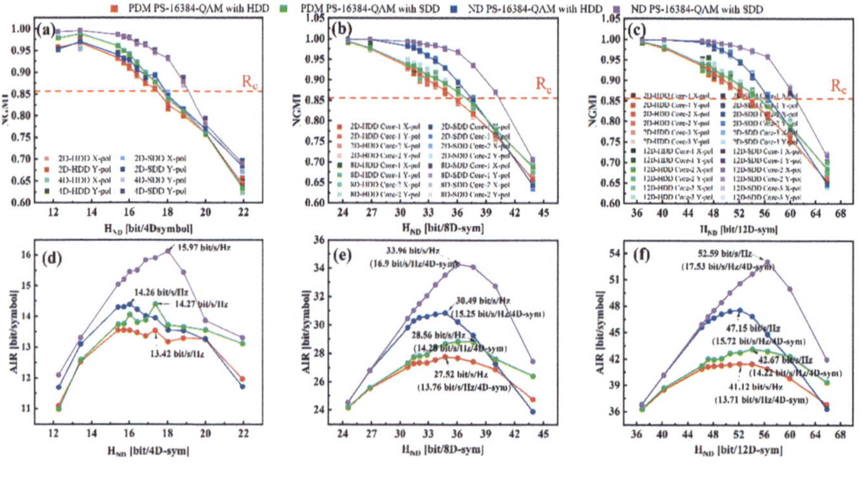

Fig. 5.27 The experimental results of NGMI for (**a**) 4D, (**b**) 8D and (**c**) 12D as a function of H_{ND} for various entropy, and AIR for (**d**) 4D, (**e**) 8D and (**f**) 12D

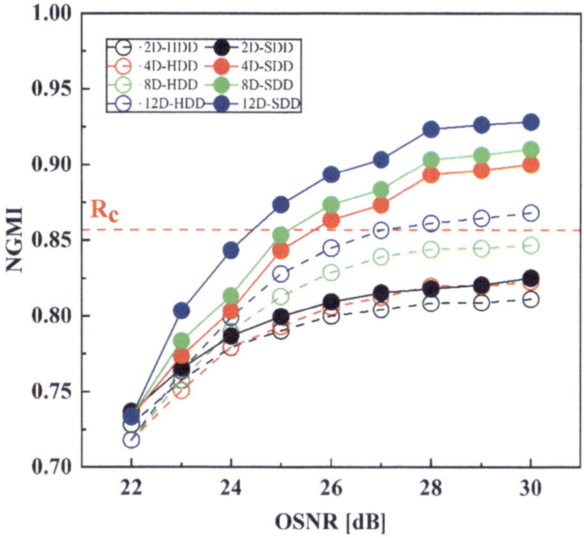

Fig. 5.28 The experimental results of NGMI for 22.5 km MCF transmission as a function of OSNR

22.5 km MCF. The maximum AIR of the 2D signals with HDD and SDD in Fig. 5.9 d at H_{ND} = 17.38 bit/4D-sym are 13.55 bit/4D-sym and 14.41 bit/4D-sym, corresponding to SEs of 13.42 bit/s/Hz and 14.27 bit/s/Hz. The maximum AIRs of the 4D signals with HDD and SDD at H_{ND} = 16.03 bits/4D-sym and 18.03 bit/4D-sym are 14.40 bits/4D-sym and 16.13 bits/4D-sym, corresponding to SEs of 14.26 bit/s/Hz and 15.97 bit/s/Hz, respectively. The gain of SE for HDD and SDD is 0.84 bit/s/Hz and 1.7 bit/s/Hz, respectively. The maximum SE of the 8D signal with HDD and SDD in Fig. 5.9 e at H_{ND} = 34.6 bit/8D-sym and 36.06 bit/8D-sym are 30.49 bit/s/Hz (15.25 bit/s/Hz/4D-sym) and 33.96 bit/s/Hz (16.9 bit/s/Hz/4D-sym), respectively. The SE gains versus 2D signals for HDD and SDD are 1.49 bit/s/Hz/4D-sym and 2.62 bit/s/Hz/4D-sym, respectively. The maximum SE at H_{ND} = 51.9 bit/12D-sym and 56.4 bit/12D-sym for the 12D signal with HDD and SDD in Fig. 5.19 f are 47.15 bit/s/Hz (15.72 bit/s/Hz/4D-sym) and 52.59 bit/s/Hz (17.53 bit/s/Hz/4D-sym), respectively. The SE gains versus 2D signals are 2.01 bit/s/Hz/4D-sym and 3.31 bit/s/Hz/4D-sym, respectively.

Figure 5.28 illustrates the NGMI as a function of OSNR for ND signal over 22.5 km MCF with H_{4D} = 18.8 bit/4D-sym. At high OSNR, it provides higher AIR with ND PS-16384-QAM compared to PDM PS-16384-QAM. SDD clearly outperforms HDD. The 4D/8D/12D-SDD signals satisfy Rc at 25.7 dB, 25.1 dB, 24.4 dB, respectively. The OSNR gains of the 8D/12DSDD are 0.6 dB and 1.3 dB, compared to 4D-SDD. 12D-HDD is satisfied at 26.9 dB, while 4D/8D-HDD and 2D-HDD/SDD do not fulfill the code rate limitation requirement. At the low SNR region, the curves tend to merge as the OSNR deteriorates. For OSNR below 23 dB, the advantage of ND-HDD disappears, even leading to a lower NGMI than 2D-HDD and 2D-SDD, but the advantage of ND-SDD remains.

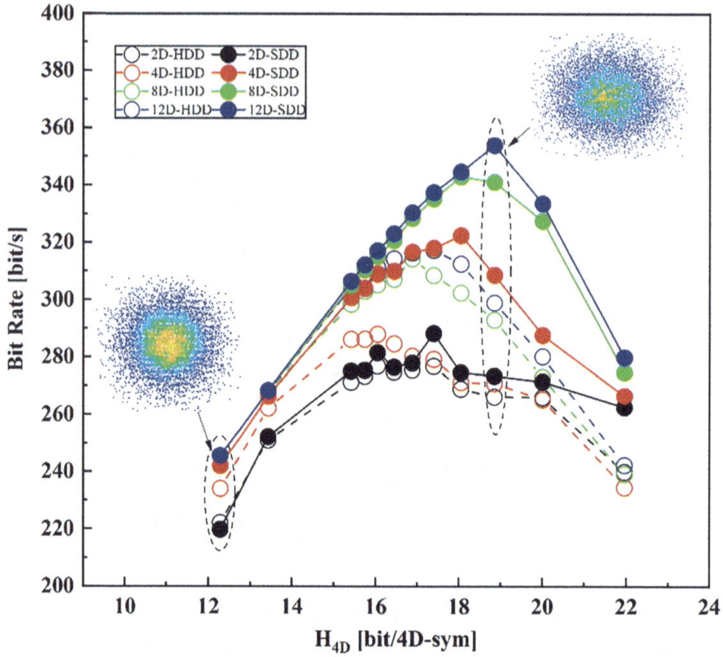

Fig. 5.29 Single-carrier bit rate for each MCF core

Figure 5.29 shows the bit rate per fiber core for ND signals as a function of H4D. The inset is the constellation. The maximum bit rate per fiber core obtained for the 4D/8D/12D-HDD signals is 288.8 Gbit/s, 314.5 Gbit/s, and 318.2 Gbit/s, which is 12 Gbit/s, 37.7 Gbit/s, and 41.4 Gbit/s more than the 2D-HDD signals, respectively. The bit rate of the 8D/12D-HDD is 25.7 Gbit/s and 29.4 Gbit/s more than that of the 4D-HDD. The maximum bit rate per fiber core obtained for the 4D/8D/12D-SDD signals is 322.6 Gbit/s, 343.2 Gbit/s, and 354.1 Gbit/s, which is 34.4 Gbit/s, 55 Gbit/s, and 65.9 Gbit/s more than the 2D-SDD signals, respectively. The bit rate of the 8D/12D-SDD is 20.6 Gbit/s and 31.5 Gbit/s more than that of the 4D-SDD. As a result, the maximum total rate that can be obtained for 12D-SDD signals in a four-core MCF-SHCD system is 1.06 Tbit/s.

5.5 Summary

In this chapter, we first discuss the fundamental principles and key technologies of SDM. Next, we propose a TA-CMA algorithm to mitigate signal impairments induced by the few-mode fiber channel, followed by a VSS-FDLMS algorithm implemented in a 7-core 6-mode optical SDM transmission system. Furthermore, we experimentally demonstrate 16-QAM signal transmission utilizing a KK receiver and employ MD modulation formats for the MCF-SHCD system.

References

1. A.A. Juarez, C.A. Bunge, S. Warm, et al., Perspectives of principal mode transmission in mode-division multiplex operation. Opt. Express **20**(13), 13810–13823 (2012)
2. M.B. Shemirani, W. Mao, R.A. Panicker, et al., Principal modes in graded-index multimode fiber in presence of spatial-and polarization-mode coupling. J. Lightwave Technol. **27**(10), 1248–1261 (2009)
3. P. Sillard, D. Molin, M. Bigot-Astruc, et al., Micro-bend-resistant low-DMGD 6-LP-mode fiber, in *Optical Fiber Communication Conference, March 20–22, 2016, Anaheim, California. Washington, D.C. OSA, 2016:Th1J.5*, (2016)
4. R.G.H. van Uden, C.M. Okonkwo, V.A.J.M. Sleiffer, H. de Waardt, A.M.J. Koonen, MIMO equalization with adaptive step size for few-mode fiber transmission systems. Opt. Express **22**, 119–126 (2014)
5. T. Mizuno, T. Kobayashi, H. Takara, A. Sano, H. Kawakami, T. Nakagawa, Y. Miyamoto, Y. Abe, T. Goh, M. Oguma, T. Sakamoto, Y. Sasaki, I. Ishida, K. Takenaga, S. Matsuo, K. Saitoh, T. Morioka, 12-core × 3-mode dense space division multiplexed transmission over 40 km employing multi-carrier signals with parallel MIMO equalization, in *Optical Fiber Communication Conference: Post deadline Papers*, (Optica Publishing Group, 2014) paper Th5B.2
6. K. Shibahara, T. Mizuno, H. Ono, K. Nakajima, Y. Miyamoto, Long-haul DMD-unmanaged 6-mode-multiplexed transmission employing cyclic mode-group permutation, in *Optical Fiber Communication Conference (OFC) 2020, OSA Technical Digest*, (Optica Publishing Group, 2020) paper Th3H.3
7. T. Zhao, F. Wen, F. Li, B. Wu, F. Fan, K. Qiu, Noise compensation for a nonlinear six-mode fiber channel through self- recycling training equalizer, in *26th Optoelectronics and Communications Conference, OSA Technical Digest*, ed. by P. Alexander Wai, H. Tam, C. Yu, (Optica Publishing Group, 2021) paper M3F.5
8. C. Wu, W. Zhang, Y. Wang, C. Gao, Study on the performance of the variable step-size LMS algorithms, in *2020 IEEE International Conference on Information Technology, Big Data and Artificial Intelligence (ICIBA), Chongqing, China, 2020*, (2020), pp. 159–162
9. R. Vijay, B. Kumar, P. Shukla, The application and simulation of block varying step size FDLMS adaptive algorithm for system identification problem, in *2012 Fourth International Conference on Computational Intelligence and Communication Networks, Mathura, India, 2012*, (2012), pp. 685–689
10. X. He, Y. Weng, Z. Pan, A step-size controlled method for fast convergent adaptive FD-LMS algorithm in few-mode fiber communication systems. J. Lightwave Technol. **32**(22), 4422–4428 (2014)
11. N. Bai, E. Ip, M. Li, T. Wang, G. Li, Experimental demonstration of adaptive frequency-domain equalization for mode-division multiplexed transmission, in *Optical Fiber Communication Conference/National Fiber Optic Engineers Conference 2013, OSA Technical Digest*, (2013)
12. D. Zou et al., 100G PAM-6 and PAM-8 signal transmission enabled by pre-chirping for 10-km Intra-DCI utilizing MZM in C band. J. Lightwave Technol. **38**(13), 3445–3453 (2020)
13. M.G. Saber et al., Demonstration of 108 Gb/s duo-binary PAM-8 transmission and the probabilistic modeling of DB-PAM-M BER. IEEE Photon. J. **13**(1), 1–14 (2021) Art no. 7200814
14. Y. Fu et al., Single-photodiode 100 Gbaud PAM-6 transmission with extended transmitter bandwidth using optical time and polarization interleaving, in *2021 Optical Fiber Communications Conference and Exhibition (OFC)*, (2021), pp. 1–3
15. J. Zhang et al., Demonstration of 260-Gb/s single-lane EML-based PS-PAM-8 IM/DD for datacenter interconnects, in *2019 Optical Fiber Communications Conference and Exhibition (OFC)*, (2019), pp. 1–3

16. A. Amari, A. Richter, Probabilistic shaping for PAM-8 optical ethernet systems, in *2020 22nd International Conference on Transparent Optical Networks (ICTON)*, (2020), pp. 1–4. https://doi.org/10.1109/ICTON51198.2020.9203547
17. J. Zhang et al., 280 Gb/s IM/DD PS-PAM-8 Transmission over 10 km SSMF at O-band for optical interconnects, in *2020 Optical Fiber Communications Conference and Exhibition (OFC)*, (2020), pp. 1–3
18. T. Wiegart, F. Da Ros, M.P. Yankov, F. Steiner, S. Gaiarin, R.D. Wesel, Probabilistically shaped 4-PAM for short-Reach IM/DD links with a peak power constraint. J. Lightwave Technol. **39**(2), 400–405 (2001). https://doi.org/10.1109/JLT.2020.3029371
19. G. Rademacher et al., Long-haul transmission over few-mode fibers with space-division multiplexing. J. Lightwave Technol. **36**(6), 1382–1388 (2018)
20. Y. Chen et al., 41.6 Tbit/s C-band SDM OFDM transmission through 12 spatial and polarization modes over 74.17 km few mode fiber. J. Lightwave Technol. **33**(7), 1440–1444 (2015)
21. D. Godard, Self-recovering equalization and carrier tracking in two-dimensional data communication systems. IEEE Trans. Commun. **28**(11), 1867–1875 (1980)
22. M.J. Ready, R.P. Gooch, Blind equalization based on radius directed adaptation. Int. Conf. Acoust. Speech Signal Process. **3**, 1699–1702 (1990)
23. Cisco, U, *Cisco Annual Internet Report (2018–2023) White Paper* (San Jose, CA, USA, Cisco, 2020)
24. F. Wang, H. Yao, J. Wang, T. Mai, X. Xin, M. Guizani, Hybrid optical-electrical data center networking: Challenges and solutions for bandwidth resource optimization. IEEE Commun. Mag. **60**, 90–96 (2022)
25. F. Chai, Q. Zhang, H. Yao, X. Xin, R. Gao, M. Guizani, Joint multi-task offloading and resource allocation for mobile edge computing systems in satellite IoT. IEEE Trans. Veh. Technol. **72**, 7783–7795 (2023)
26. D.J. Richardson, J.M. Fini, L.E. Nelson, Space-division multiplexing in optical fibres. Nat. Photon. **7**, 354–362 (2013)
27. G.M. Saridis, D. Alexandropoulos, G. Zervas, D. Simeonidou, Survey and evaluation of space division multiplexing: From technologies to optical networks. IEEE Commun. Surv. Tutor **17**, 2136–2156 (2015)
28. B.J. Puttnam, G. Rademacher, R.S. Luís, Space-division multiplexing for optical fiber communications. Optica **8**, 1186–1203 (2021)
29. S. Zhou, X. Liu, R. Gao, Z. Jiang, H. Zhang, X. Xin, Adaptive Bayesian neural networks nonlinear equalizer in a 300-Gbit/s PAM8 transmission for IM/DD OAM mode division multiplexing. Opt. Lett. **48**, 464–467 (2023)
30. L. Zhu, H. Yao, H. Chang, Q. Tian, Q. Zhang, X. Xin, F.R. Yu, Adaptive Optics for orbital angular momentum-based internet of underwater things applications. IEEE Internet Things J. **9**, 24281–24299 (2022)
31. J. Zhang, J. Liu, Z. Lin, J. Liu, L. Shen, S. Yu, Nonlinearity-aware adaptive bit and power loading DMT transmission over low-crosstalk ring-core fiber with mode group multiplexing. J. Lightwave Technol. **38**, 5875–5882 (2020)
32. H. Tan, J. Deng, R. Zhao, X. Wu, G. Li, L. Huang, J. Liu, X. Cai, A free-space orbital angular momentum multiplexing communication system based on a metasurface. Laser Photon. Rev. **13**, 1800278 (2019)
33. J. Liu, G. Zhu, J. Zhang, Y. Wen, X. Wu, Y. Zhang, Y. Chen, X. Cai, Z. Li, Z. Hu, et al., Mode division multiplexing based on ring core optical fibers. IEEE J. Quant. Electron. **54**, 1–18 (2018)
34. G. Zhu, Z. Hu, X. Wu, C. Du, W. Luo, Y. Chen, X. Cai, J. Liu, J. Zhu, S. Yu, Scalable mode division multiplexed transmission over a 10-km ring-core fiber using high-order orbital angular momentum modes. Opt. Express **26**, 594–604 (2018)
35. B. Zhu, SDM fibers for data center applications, in *Proceedings of the 2019 Optical Fiber Communications Conference and Exhibition (OFC), San Diego, CA, USA, 3–7 March 2019*, (2019), p. M1F.4

36. D.L. Butler, M.J. Li, S. Li, Y. Geng, R.R. Khrapko, R.A. Modavis, V.N. Nazarov, A.V. Koklyushkin, Space division multiplexing in short reach optical interconnects. J. Lightwave Technol. **35**, 677–682 (2016)
37. D. Qian, E. Ip, M.F. Huang, M.J. Li, A. Dogariu, S. Zhang, Y. Shao, Y.K. Huang, Y. Zhang, X. Cheng, et al., 105 Pb/s Transmission with 109 b/s/Hz spectral efficiency using hybrid single- and few-mode cores, in *Proceedings of the Frontiers in Optics, Rochester, NY, USA, 14–-18 October 2012*, (2012), p. FW6C.3
38. D. Kong, A. Jørgensen, M. Henriksen, F. Klejs, Z. Ye, Ò. Helgason, H. Hansen, H. Hu, M. Yankov, S. Forchhammer, et al., Single dark-pulse kerr comb supporting 1.84 Pbit/s transmission over 37-core fiber, in *Proceedings of the CLEO: QELS_Fundamental Science, Washington, DC, USA, 10–15 May 2020*, (2020), p. JTh4A.7
39. T. Hayashi, T. Nagashima, T. Morishima, Y. Saito, T. Nakanishi, Multi-core fibers for data center applications, in *Proceedings of the 45th European Conference on Optical Communication (ECOC 2019), Dublin, Germany, 22–26 September 2019*, (2019), pp. 1–4
40. S.T. Le, K. Schuh, H.N. Tan, A closed-form expression for direct detection transmission systems with Kramers-Kronig receiver. IEEE Photon. Technol. Lett. **30**, 2048–2051 (2018)
41. K. Zhong, X. Zhou, J. Huo, C. Yu, C. Lu, A.P.T. Lau, Digital signal processing for short-reach optical communications: A review of current technologies and future trends. J. Lightwave Technol. **36**, 377–400 (2018)
42. A. Mecozzi, C. Antonelli, M. Shtaif, Kramers–Kronig coherent receiver. Optica **3**, 1220–1227 (2016)
43. Y. Liu, Y. Li, J. Song, L. Yue, H. Zhou, M. Luo, Z. He, J. Qiu, Y. Zuo, W. Li, et al., Transmission of a 112-Gbit/s 16-QAM over a 1440-km SSMF with parallel enabled by an overlap approach and bandwidth compensation. Opt. Express **29**, 8117–8129 (2021)
44. A. Mecozzi, C. Antonelli, M. Shtaif, Kramers–Kronig receivers. Adv. Opt. Photon. **11**, 480–517 (2019)
45. R.S. Luís, G. Rademacher, B.J. Puttnam, Y. Awaji, N. Wada, Long distance crosstalk-supported transmission using homogeneous multicore fibers and SDM-MIMO demultiplexing. Opt. Express **26**, 24044–24053 (2018)
46. N.K. Fontaine, C.R. Doerr, M.A. Mestre, R.R. Ryf, P.J. Winzer, L.L. Buhl, Y. Sun, X. Jiang, R. Lingle, Space-division multiplexing and all-optical MIMO demultiplexing using a photonic integrated circuit, in *Proceedings of the OFC/NFOEC, Los Angeles, CA, USA, 4–6 March 2012*, (2012), pp. 1–3
47. S. Randel, S. Corteselli, D. Badini, D. Pilori, S. Caelles, S. Chandrasekhar, J. Gripp, H. Chen, N.K. Fontaine, R. Ryf, et al., First real-time coherent MIMO-DSP for six coupled mode transmission, in *Proceedings of the 2015 IEEE Photonics Conference (IPC), Reston, VA, USA, 4–8 October 2015*, (2015), pp. 1–2
48. S. Beppu, K. Igarashi, M. Kikuta, D. Soma, T. Nagai, Y. Saito, H. Takahashi, T. Tsuritani, I. Morita, M. Suzuki, Weakly coupled 10-mode-division multiplexed transmission over 48-km few-mode fibers with real-time coherent MIMO receivers. Opt. Express **28**, 19655–19668 (2020)
49. R.S. Luis, B.J. Puttnam, G. Rademacher, Y. Awaji, N. Wada, On the use of high-order MIMO for long-distance homogeneous single-mode multicore fiber transmission, in *Proceedings of the 2017 European Conference on Optical Communication (ECOC), Gothenburg, Sweden, 17–-21 September 2017*, (2017), pp. 1–3
50. K. Shibahara, T. Mizuno, D. Lee, Y. Miyamoto, Advanced MIMO signal processing techniques enabling long-haul dense SDM transmissions. J. Lightwave Technol. **36**, 336–348 (2017)
51. X. Chen, C. Antonelli, S. Chandrasekhar, G. Raybon, A. Mecozzi, M. Shtaif, P. Winzer, Kramers–Kronig receivers for 100-km datacenter interconnects. J. Lightwave Technol. **36**, 79–89 (2018)
52. A. Mecozzi, A necessary and sufficient condition for minimum phase and implications for phase retrieval. arXiv 2016, arXiv:1606.04861

53. C. Antonelli, M. Shtaif, A. Mecozzi, Modeling of nonlinear propagation in space-division multiplexed fiber-optic transmission. J. Lightwave Technol. **34**, 36–54 (2015)
54. X. Chen, S. Chandrasekhar, S. Randel, W. Gu, P. Winzer, Experimental quantification of implementation penalties from limited ADC resolution for Nyquist shaped higher-order QAM, in *Proceedings of the 2016 Optical Fiber Communications Conference and Exhibition (OFC), Anaheim, CA, USA, 20–24 March 2016*, (2016), p. W4A.3
55. J. Zhang, J. Yu, X. Li, Y. Wei, K. Wang, L. Zhao, W. Zhou, M. Kong, X. Pan, B. Liu, et al., 100 Gbit/s VSB-PAM-n IM/DD transmission system based on 10 GHz DML with optical filtering and joint nonlinear equalization. Opt. Express **27**(5), 6098–6105 (2019)
56. V.J. Mathews, J. Lee, A fast recursive least-squares second order Volterra filter, in *Proceedings of the ICASSP-88, International Conference on Acoustics, Speech, and Signal Processing, New York, NY, USA, 11–14 April 1988*, (1988), pp. 1383–1384
57. Y. Gao, X. Zhou, F. Li, J. Huo, J. Yuan, K. Long, Mismatch length estimation of self-homodyne coherent optical systems by using carrier-pilot-assist method, in *2022 Optical Fiber Communications Conference and Exhibition (OFC), San Diego, CA, USA, 2022*, (2022), pp. 01–03
58. K. Gallo, G. Assanto, All-optical diode based on second-harmonic generation in an asymmetric waveguide. J. Opt. Soc. Am. B **16**(2), 267–269 (1999)
59. X. Zhou, R. Urata, H. Liu, Beyond 1 Tb/s intra-data center interconnect technology: IM-DD OR coherent? J. Lightwave Technol. **38**(2), 475–484 (2020). https://doi.org/10.1109/JLT.2019.2956779
60. R. Zhang, K. Kuzmin, Y.-W. Chen, W.I. Way, 800G/λ self-homodyne coherent links with simplified DSP for next-generation intra-data centers. J. Lightwave Technol. **41**(4), 1216–1222 (2023). https://doi.org/10.1109/JLT.2022.3218764
61. D. Tauber et al., Role of coherent systems in the next DCI generation. J. Lightwave Technol. **41**(4), 1139–1151 (2023). https://doi.org/10.1109/JLT.2023.3235820
62. B.J. Puttnam et al., High-capacity transmission systems using homogeneous multi-core fibers. J. Lightwave Technol. **35**(6), 1157–1167 (2017). https://doi.org/10.1109/JLT.2017.2669207
63. Y. Koizumi, K. Toyoda, M. Yoshida, M. Nakazawa, 1024 QAM (60 Gbit/s) single-carrier coherent optical transmission over 150 km. Opt. Express **20**(11), 12508–12514 (2012)
64. S. Beppu, K. Kasai, M. Yoshida, M. Nakazawa, 2048 QAM (66 Gbit/s) single-carrier coherent optical transmission over 150 km with a potential SE of 15.3 bit/s/Hz. Opt. Express **23**(4), 4960–4969 (2015)
65. X. Chen, S. Chandrasekhar, J. Cho, P. Winzer, Transmission of 30-GBd polarization-multiplexed probabilistically shaped 4096-QAM over 50.9-km SSMF. Opt. Express **27**(21), 29916–29923 (2019)
66. X. Chen, J. Cho, A. Adamiecki, P. Winzer, 16384-QAM transmission at 10 GBd over 25-km SSMF using polarization-multiplexed probabilistic constellation shaping, in *45th European Conference on Optical Communication (ECOC 2019), Dublin, Ireland, 2019*, (2019), pp. 1–4. https://doi.org/10.1049/cp.2019.1030
67. J. Ding et al., Demonstration of 352-Gbit/s single line rate PS-4096QAM THz wired transmission over hollow-core fiber, in *2021 Opto-Electronics and Communications Conference (OECC), Hong Kong, Hong Kong, 2021*, (IEEE, 2021), pp. 1–3

Chapter 6
Channel and Equalization Algorithms in Optical Fiber Communication System

6.1 Equalization and Compensation Algorithm

6.1.1 MIMO Algorithm for SDM Channel Impairments

6.1.1.1 Joint Algorithm Composed of CMA and DDLMS

With the rising popularity of internet services, it is necessary to increase the capacity of fiber optical transmission [1–3]. To overcome the limit of standard single mode fiber (SSMF), space-division multiplexed (SDM) transmission based on multi-core (MCF) or multi-mode fiber (MMF) is one of the most popular solutions nowadays [4, 5]. The capacity of SDM transmission can be promoted by times with the increase of the cores and the spatial modes [6].

Recently, several groups demonstrated the feasibility of applying mode-division multiplexing (MDM) and multiple-input multiple-output (MIMO) in few-mode fiber (FMF) transmission [7–10]. In [7], a MIMO-less 2 × 25-Gb/s OOK signal and 2 × 20-Gb/s PAM-4 signal transmission over 15-km FMF have been demonstrated. In [8], An Li et al. focused on the performance of MMF and FMF transmission with a data rate of 100-Gb/s respectively.

In this section, we experimentally demonstrate a 2 × 20G Baud PAM4 Transmission over 20 km. The joint algorithm is proposed which is composed of CMA and DDLMS. The result shows that the BER can be reduced to 1.0×10^{-3} with the joint composed of CMA and DDLMS while the received power is 0 dB. We also discussed the effect of symbol rate and transmission distance on the performance.

Figure 6.1 shows the experimental setup. At the transmitter, the light is emitted from an external cavity (ECL) at 1550 nm with 100 kHz linewidth. The electronic signal generated from an arbitrary waveform generator with 65 GSa/s was modulated by the 40GHz Mach–Zehnder modulator (MZM) to achieve electrical/optical conversion. Then the erbium-doped fiber amplifier (EDFA) is used to provide the power gain into the 1 × 6 optical coupler. The delay line is used to resolve the correlation between modes.

F. Tian, X. Xin, *High-Speed Large Capacity Optical Fiber Communications*, Wireless Networks, https://doi.org/10.1007/978-3-031-94211-2_6

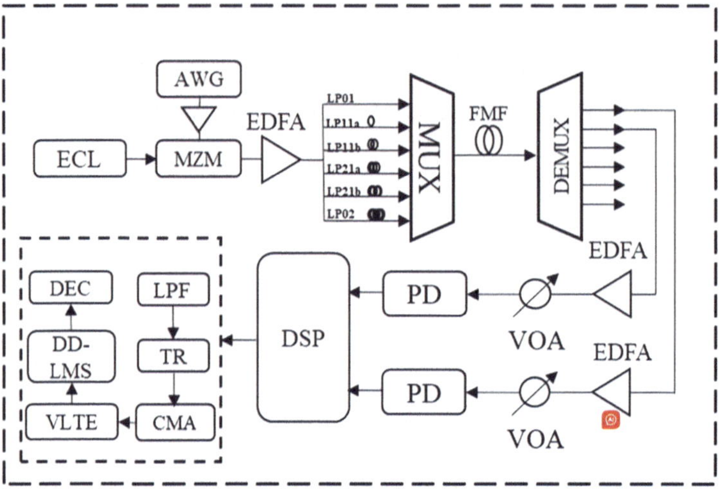

Fig. 6.1 Experiment setup for the 2 × 20G baud/s PAM4 system

Table 6.1 Insertion loss of each mode

Modes	LP01	LP11a	LP11b	LP21a	LP21b	LP02
MUX1	1.44	1.48	1.6	2	3.05	5.19
MUX2	1.25	0.9	0.92	1.88	1.89	4.37

Afterwards, the signals are coupled into the six-mode spatial-multiplexer (MUX). The transmission link consists of two unrepeated spans of 10 km FMF and then we used a demultiplexer (DEMUX) to spilt the signals. The EDFA and a variable optical attenuator are used for adjusting the launched signal power into the 50 GHz photodetector. Then a digital phosphor oscilloscope (Tektronix DPO72004C) with 100 GSa/s and 20 GHz bandwidth were used to receive the signal. We then processed the digitized signal offline using MATLAB.

The offline DSP is performed at receiver which includes low pass filter, time recovery, cascaded modulus algorithm (CMA), decision-directed least mean square (DD-LMS) algorithm and the BER calculation.

Table 6.1 shows the insertion loss of each mode. It can be seen that, with the increasing of the mode-order, the insertion loss also has tendency of elevation. The insertion loss of LP01 and LP21 are 1.44 dB and 5.19 dB respectively.

Figure 6.2 shows the beam intensity distribution map after 10 km and 20 km respectively. It can be seen that the LP01 mode did not deteriorate after 10 km and 20 km transmission. The orthogonality between degenerate modes will also get worse as the transmission distance increases.

Figure 6.3 shows the measured bit error rate (BER) curves versus received power in the 10 km 20G baud/s PAM4 system. It is worth noting that BER decreases as the received optical power increases after 10 km transmission. With the order of mode increasing, the transmission performance will also decrease. For degenerate modes, the transmission performance shows a similar trend. Take Received Power being

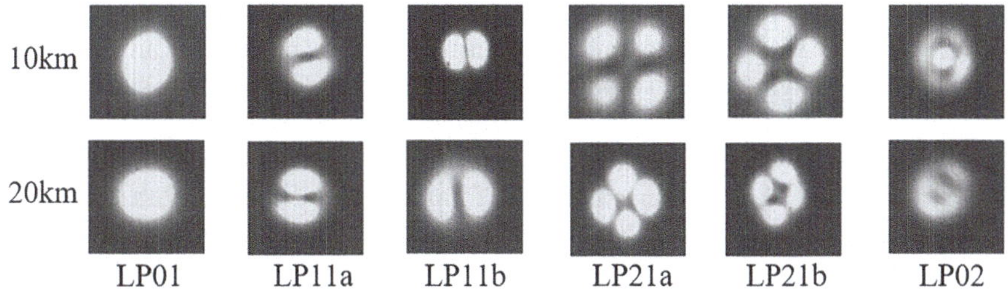

Fig. 6.2 The beam intensity distribution map after 10 km and 20 km

Fig. 6.3 Measured BER curves versus received power in the 10 km 20G baud/s PAM4

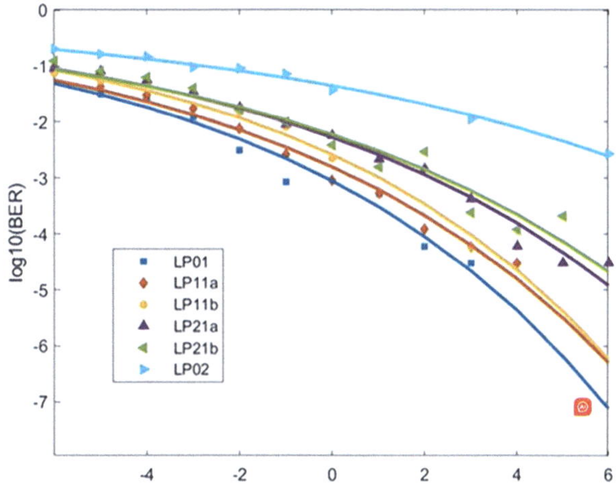

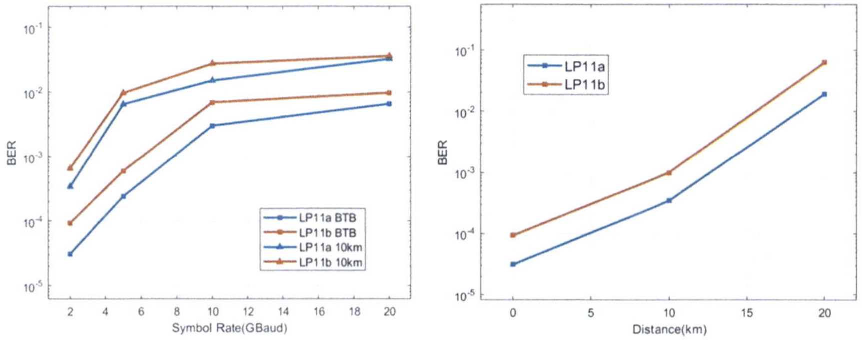

Fig. 6.4 Results of 2 × 20G baud/s PAM4 system: (**a**) Measured BER curves versus symbol rate. (**b**) Measured BER curves versus distance

0 dB as an example, the BER of LP01 is 1.92×10^{-3}, and the BER of LP11b and LP21b are 3.27×10^{-3} and 5.83×10^{-3}.

Figure 6.4a demonstrated the measured BER curves versus Symbol Rate from 2GBaud to 20GBaud. It is worth noting that, when the transmission symbol rate increases, the transmission quality shows some differences accordingly. With the signal rate increasing, the crosstalk between the various modes has an increased influence on the signal. LP11a mode will have better transmission performance than the mode of LP11b due to the performance difference of the MUX and DEMUX. We can see that BER and symbol rate have a positive correlation. Figure 6.4b demonstrated the relationship of transmission performance and distance. With the transmission distance increasing, the crosstalk between degenerate modes will also increase, which will lead to the quality deterioration of the received signal.

In this section, we experimentally demonstrated a 2 × 20G Baud PAM4 Transmission over 20 km. The joint algorithm, composed of CMA and DDLMS was proposed. The result shows that the BER can be reduced to 1.0×10^{-3} with the joint composed of CMA and DDLMS while the received power is 0 dB. We also discussed the effect of symbol rate and transmission distance on the performance.

6.1.2 SCA-CMMA-Based MIMO Equalization Algorithm

With the gaining popularity of large-data-volume Internet services such as cloud computing, 4 K video, AR, and VR, optical fiber communication systems are facing tremendous capacity growth pressure. The space division multiplexing technology is used to multiplex information in the signal space dimension, which can greatly increase the capacity of optical fiber communication systems [11, 12]. The multi-core, low-mode fiber-based space division multiplexed optical transmission system not only breaks the experimental record of the high transmissions data rate in the single-mode optical fiber communication system, but also shows applicability in different scenarios such as access. In low-mode multi-core fiber transmission systems, optical signals are affected by white Gaussian noise, nonlinearity, mode coupling, and mode dispersion. Although the use of multiple-output (MIMO) signal processing has contributed significantly to the development of MIMO-SDM transmission, differential mode delay (DMD) and mode-dependent loss (MDL) remain the limiting factors for long-range MIMO SDM Transmission [13, 14]. Therefore, in order to improve both the transmission distance and performance of multi-core low-mode fiber, it is of special interest to study the equalization algorithm at the signal's receiving end [15]. Based on the seven-core three-mode optical transmission model, we propose a new improved algorithm based on the Square Contour Algorithm (SCA) [16]. This algorithm is a combination of the SCA and CMMA algorithms [17]. The two algorithms are switched by setting up a decision domain. This SCA-CMMA joint equalization algorithm can accurately equalize differential mode group delay (DMGD) and mode dependent loss (MDL), which demonstrate a better performance with faster convergence speed and lower steady state error.

Among the algorithms on the receiving end of a multi-core, few-mode optical transmission link, except for the MIMO equalization part [18]. Other technologies such as dispersion compensation, frequency offset estimation, and carrier phase

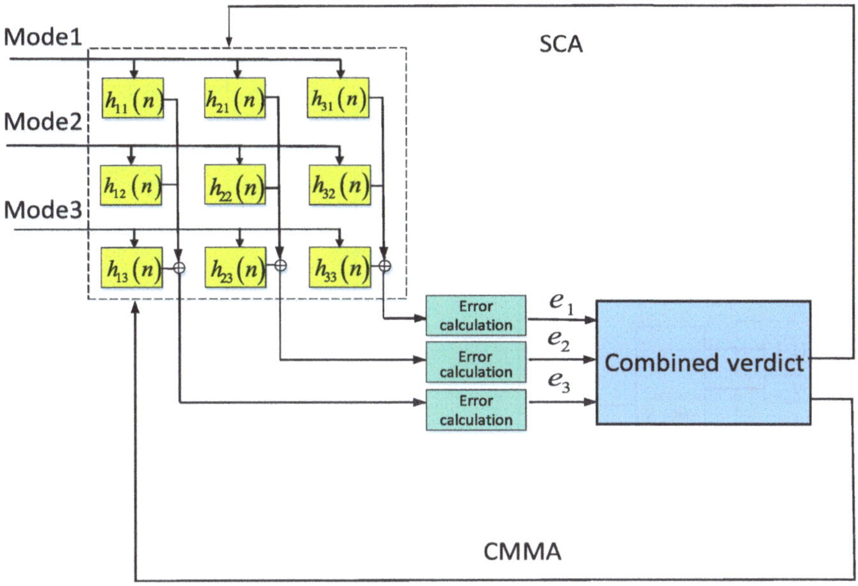

Fig. 6.5 SCA-CMMA Algorithm decision flow chart

recovery are the same as those used in a standard single-mode fiber. The main research center of high-end algorithms is mainly on MIMO equalization.

This part focuses on improving the error performance of the receiver system. By improving the equalization effect of the MIMO equalizer, that is, speeding up the convergence speed of the algorithm and reducing the steady state error of the equalizer, the error performance of the communication system is thus improved. The traditional SCA has fast convergence speed, but the steady-state error is enormous. The CMMA has a slower convergence speed, but its steady-state error effect is better than that of SCA. Therefore, we use SCA as the cold-start algorithm of CMMA, so that the result of equalization can have the advantages of being both algorithms. We apply the proposed new algorithm at the receiver-side MIMO algorithm in the established seven-core three-mode optical transmission model. In multi-core and low-mode fiber, the inter-core crosstalk can be controlled when designing the fiber, so MIMO can be avoided in the receiver algorithm. We put the SCA-CMMA algorithm in 3×3 equalizer Equalization-mode-related impairments. The specific flowchart of this algorithm is shown in Fig. 6.5. The optical signals of three modes in each core passed through a MIMO equalizer, and the equalization process is performed. The error is calculated by comparing the output results with the standard constellation points. The error value passes through the joint decision device and it will enter the CMMA algorithm if the decision condition is met, otherwise it continues to use the SCA algorithm to calculate the value of the tap coefficient in the MIMO equalizer. This decision threshold is obtained through

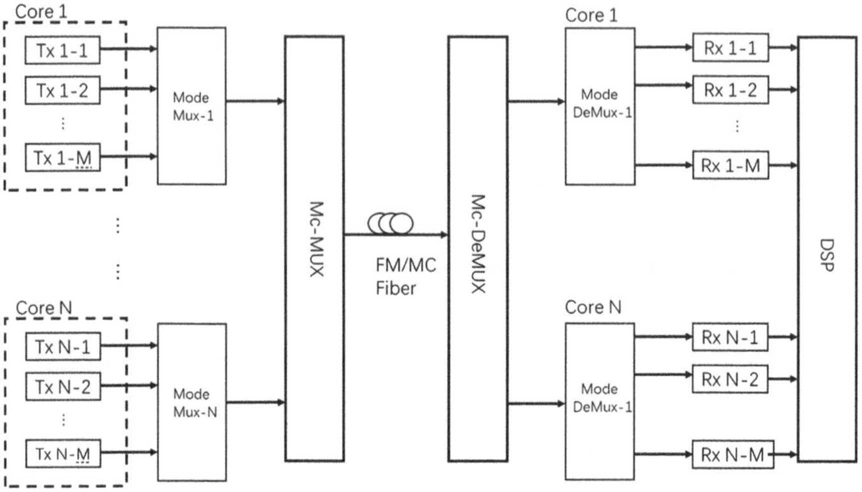

Fig. 6.6 System setup (FM/MC: few mode/multi-core, DSP: digital signal processing)

multiple experiments. We take $e1 = 0.3$, $e2 = 0.7$, and $e3 = 0.7$ as the decision threshold.

The system setup is shown in Fig. 6.6. At the transmitter, twenty-one 16QAM modulated sequences are separated into seven groups, which are sent into different cores. In each core, there are three modes including LP01, LP11a, and LP11b. After 50 km FM/MC fiber transmission, the signal is received by multi-mode coherent receiver. At the receiver side, the signal is firstly separated into seven groups by the multi-core DeMux and then detected by the multi-mode coherent receiver. After analog to digital convertor, the 21 digital streams are equalized by our proposed SCA-CMMA MIMO algorithm.

The traditional SCA algorithm has a fast convergence speed, yet a high steady-state error. CMMA algorithm has slow convergence speed, but relatively low steady-state error. The improved algorithm has the advantages of both, which can guarantee both fast convergence and lower steady-state error. Figure 6.7 illustrates the measured system performance. Compared with traditional MIMO equalization, our proposed method improves the optical signal-to-noise ratio (OSNR) with a BER of FEC by about 3 dB. Since MIMO equalization based on SCA-CMMA can select an algorithm with a better equalization result according to the error function, it can achieve better system performance.

We have proposed and demonstrated SCA-CMMA-based MIMO equalization method. This method can effectively improve the transmission performance of the MIMO system without increasing the complexity of the algorithm [19]. This proposed method improves the OSNR with a BER of FEC by about 3 dB. The results show that it is a promising method for massive MIMO equalization applications, especially in FM / MC fiber.

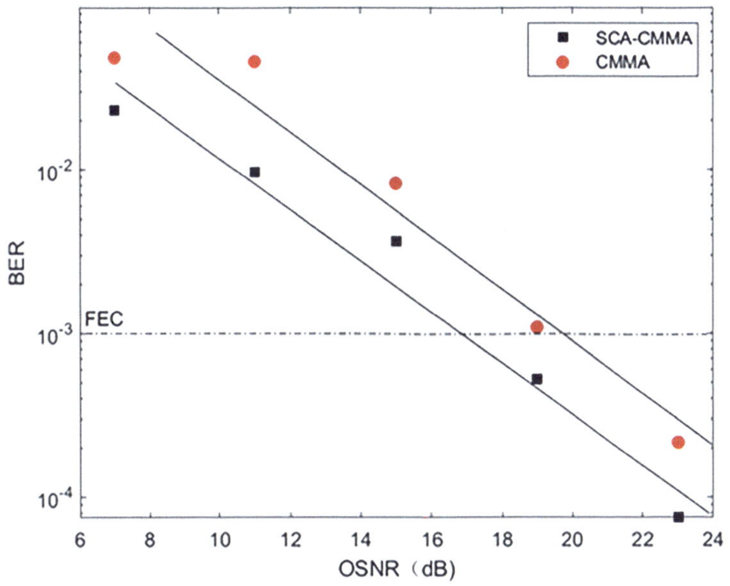

Fig. 6.7 BER performance of SCA-CMMA

6.2 Jones-Volterra Training Sequence (JVTS) Algorithm for Nonlinear Compensation

Nowadays, a huge demand on the bandwidth of optical transports has occurred by the fast growth of bandwidth-rich internet applications such as 4G-TV and industrial internet. Space-division multiplexing (SDM) technologies based on multi-core fiber (MCF) and few-mode fiber (FMF) have been widely studied for overcoming the capacity limitations of traditional single-mode fibers and increasing the transmission capacity of backbone network [20, 21]. A MIMO-less 2 × 25-Gb/s OOK signal and 2 × 20-Gb/s PAM-4 signal transmission over 15-km FMF have been demonstrated in [6]. Georg Rademacher et al. focused on large-capacity transmission with 30 km FMF respectively in [3].

Digital signal processing (DSP) receives great attention recent years in solving the signal impairment along the fiber link [1–3]. To combat the optical damage in FMF optical transmission systems, training sequence (TS) and Volterra series transfer functions (VSNF) techniques become an indispensable component of DSP [22–24]. TS has been proposed to release the dependency on the priori channel model and estimate the dynamic environment of transmission link. To alleviate the noise impact, the periodic structure of the training sequence allows for the processing of multiple period auto correlation in optical fiber transmission. In [22], Ying Zhao et al. verified a periodic training sequence structure in 16 quadratic-amplitude modulation (QAM) system. To mitigate the device nonlinearities in optical transmission systems, VSNF have also been proposed. Nonlinearities are

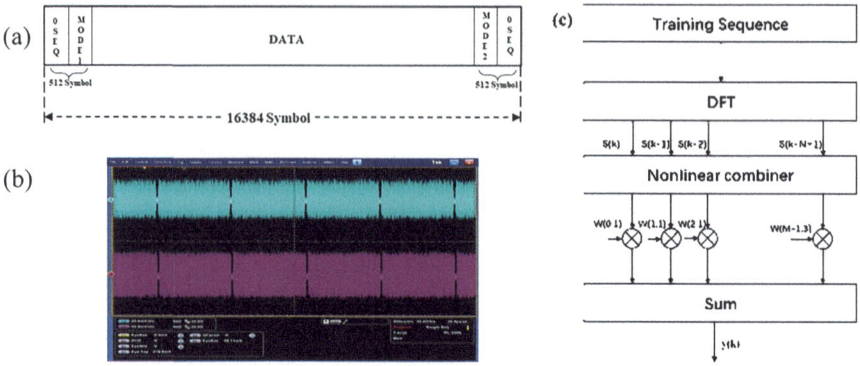

Fig. 6.8 (**a**) Training sequence structure in 2 × 2 FMF Transmission; (**b**) periodic time domain diagram of training sequence structure in oscilloscope; (**c**) third-order frequency domain Volterra Equalizer structure diagram aid by training sequence

modeled and compensated in time domain or in frequency domain by using a set of nth order nonlinear transfer functions [23]. V. Vgenopoulou et al. demonstrated a novel Volterra-based nonlinear equalizer and revealed a ~ 0.8 dB Q-factor improvement after transmitting 6 × 32Gbaud PM-16QAM signals over 1040 km [24].

In this section, we experimentally demonstrated a 2 × 20G Baud PAM4 transmission and proposed a Jones-Volterra training sequence to mitigate the device nonlinearities and dynamic crosstalk in FMF system. The results show that Jones-Volterra training sequence (JVTS) can achieve a 0.5 dB received power improvement at forward error correction coding (FEC) threshold (3.8×10^{-3}). With the help of the proposed algorithm, BER of LP01 and LP11a can be reduced to 7.16×10^{-4} and 5.8×10^{-3} while the received power to 5 dBm.

6.2.1 Analysis of JVTS

Figure 6.8a shows the training sequence structure in 2 × 2 FMF transmission. The number of symbols in one period is set at 16,384, among which 1024 symbols are taken as the training sequence. In allusion to 2 × 2 FMF transmission, we set 512 symbols each for mode 1 and mode 2 in the beginning and at the end of one period. The symbol value is set to 0 in the front 256 bits of each mode, and the last 256 bits are set to a known sequence to calculate the Jones matrix. The Jones matrix describes the mapping component of each mode, which can be described as:

$$H = \begin{bmatrix} hxx & hxy \\ hyx & hyy \end{bmatrix} \tag{6.1}$$

Where x and y represent the sequence number of the mode. For instance, *hxy*represents the mapping component of mode x to mode y. By inverting the Jones matrix, we can recover the dynamic signal impairment in FMF optical transmission systems.The periodic time domain diagram of training sequence structure in oscilloscope is shown in Fig. 6.8b. The oscilloscope shows 200,000 data points for LP01 and LP11a. Four complete periods can be observed in the oscilloscope for a triple up-sampling at the transmitter. It is worth noting that we can roughly estimate the crosstalk of the FMF optical transmission systems by observing the TS at the receiver. Figure 6.8c shows the third-order frequency domain Volterra equalizer structure diagram aid by training sequence. We obtained the Volterra polynomial coefficients by decomposing the nonlinear Schrodinger equation. The nonlinear damage caused by the device can be compensated with the aid of the training sequence. Through the joint training sequence of Jones matrix and Volterra-based nonlinear equalizer, the device nonlinearities and dynamic crosstalk could be effectively mitigated in FMF system.

6.2.2 Experimental Setup

Figure 6.9 shows the experimental setup. At the transmitter, the light is emitted from a 100 kHz linewidth external cavity (ECL) at 1550 nm. The electronic signal generated from an arbitrary waveform generator with 65 GSa/s and modulated by the 40 GHz Mach–Zehnder modulator (MZM). Then the erbium-doped fiber amplifier (EDFA) is used to provide the power gain into the 1 × 6 optical coupler. The delay line is used to resolve the correlation between modes. Afterwards, the signals are coupled into the six-mode spatial-multiplexer (MUX). The transmission link consists of 2 unrepeated spans of 10 km FMF and a demultiplexer (DEMUX) is used to spilt the signals. The EDFA and variable optical attenuator is used for adjusting the launched signal power into the 50 GHz photodetector. Then a digital phosphor

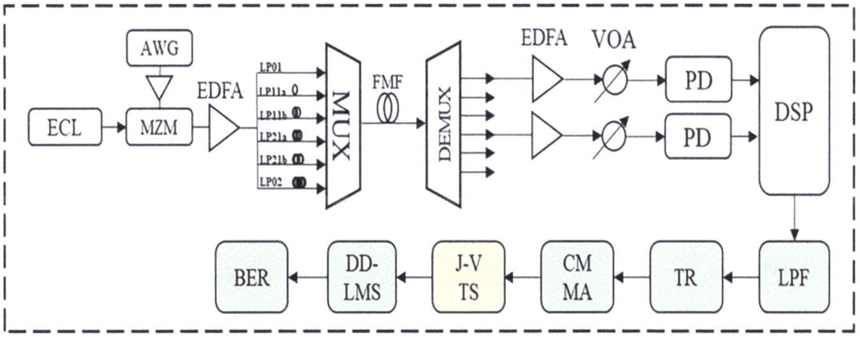

Fig. 6.9 Experiment Setup for the 2 × 20G baud/s PAM4 system

oscilloscope (Tektronix DPO72004C) with 100 GSa/s and 20 GHz bandwidth were used to receive the signal.

The offline DSP is performed at the receiver which includes low pass filter, time recovery, cascaded multi-modulus algorithm (CMMA), J-VTS, decision-directed least mean square (DD-LMS) algorithm and the BER calculation.

6.2.3 Results and Discussions

Figure 6.10a shows the result of LP01 mode and LP11a with JVTS equalization. It can be seen that JVTS has a superior equalization effect on the FMF system, and the LP01 mode has better performance, which is also confirmed in Fig. 6.10b. Figure 6.10b measured BER curves versus received power in the 2 × 20G baud/s PAM4 system. It is worth noting that BER and the received optical power show a negative correlation tendency. LP01 mode will have a better transmission performance due to the performance difference of the MUX and DEMUX and the crosstalk between the various modes. The results also show that Jones-Volterra training sequence can mitigate the device nonlinearities and dynamic crosstalk in FMF system effectively. With the help of the proposed algorithm, BER of LP01 and LP11a can be reduced to 7.16×10^{-4} and 5.8×10^{-3} while the received power is 5 dBm. At the forward error correction coding (FEC) threshold (3.8×10^{-3}), Jones-Volterra training sequence can achieve a 0.5 dB received power improvement.

In this section, we experimentally demonstrated a 2 × 20G Baud PAM4 FMF transmission and proposed a Jones-Volterra training sequence to mitigate the device nonlinearities and dynamic crosstalk. The results show that Jones-Volterra training sequence can achieve a 0.5 dB received power improvement at forward error correction coding (FEC) threshold (3.8×10^{-3}). With the help of the proposed algorithm, BER of LP01 and LP11a can be reduced to 7.16×10^{-4} and 5.8×10^{-3} when the received power is 5 dBm.

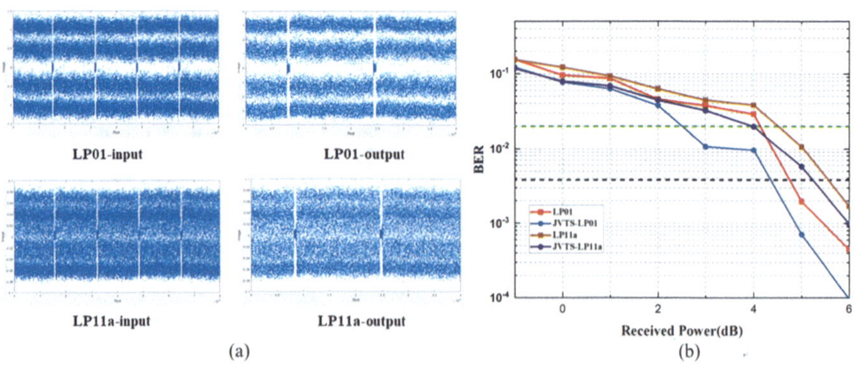

Fig. 6.10 (**a**) Scatter plot of PAM4 before JVTS and post JVTS; (**b**) measured BER curves versus received power in the 2 × 20G baud/s PAM4 system

6.3 Carrier Phase Recovery Algorithm

6.3.1 *A Feed-Forward CPR Algorithm with the Minimum FOD Approximation*

With the rapid development of high-definition video, high-speed internet, cloud computing and other advanced technologies, the requirements for large capacity in backbone networks is increasing. High-speed coherent optical communication system based on high-order modulation and advanced digital signal processing (DSP) technology has attracted much attention [25–31]. However, the tolerance towards laser phase noise decreases dramatically due to the inherent shorter Euclidean distance for M-QAM modulation formats [32]. Therefore, it is necessary to use the phase noise estimation exactly at the receiver side with the DSP technology [33, 34].

The carrier phase recovery algorithm can be classified into two categories: backward feedback methods based on feedback delay and feed-forward methods based on blind phase search (BPS). In the feedback algorithm, the feedback delay is significant and the efficiency of the decision-directed feedback method in the tolerance of linewidth is low [35, 36]. Feed forward algorithm has been widely used in practical parallel and pipeline architectures [37–46]. In 1983, Viterbi A. took the fourth power of QPSK signal and eliminated the phase information to obtain phase noise. The algorithm is named VVPE, which is not suitable for high-order modulated signal [47]. Pfau et al. proposed BPS algorithm [48], which is realized by rotating multiple test phases for the original signal and finds the best test phase as phase noise. This algorithm is suitable for M-QAM signal and the required number of test phases increases with M. In order to reduce the complexity of BPS algorithm, several multi-level feedforward algorithms based on BPS have been proposed. These algorithms reduce the complexity by reducing the number of test phases. For example, in the first stage, the BPS algorithm rotates less test phase and in the second stage, it uses the maximum likelihood (ML) estimation algorithm [49]. Bilal S M and others have proposed a series of multistage carrier phase recovery algorithms for 64QAM and 256QAM signals [50–54]. Xiang M et al. proposed carrier phase estimation (CPE) based on a two-stage BPS with quadratic approximation (QA), which can significantly accelerate the speed of phase searching [55]. Rottenberg et al. proposed a phase recovery method based on two estimators consisting of a ML and maximum a posteriori (MAP) criteria. An analytical solution is found for the phase error to avoid using multiple phase tests by linearization of the error [56]. Compared with the feedback phase recovery algorithm, the feed-forward phase recovery algorithm can process signals in parallel, and it is still a research hotspot in modern digital signal processing.

In this section, we have proposed a feed-forward carrier phase estimation algorithm based on two-stage BPS with the minimum first order distance (FOD) approximation. The proposed algorithm uses the FOD instead of blindly searching the phase with fixed step-size to reduce the complexity of the algorithm. The theoretical and experimental results show that the proposed algorithm reduces the complexity and ensures the accuracy as well.

This section introduces the principle and experimental results of our proposed algorithm. Firstly, a phase noise model was proposed: at the receiver, the received symbol y_k contains frequency offset, phase noise and spontaneous radiation noise shown by the equation below:

$$y_k = A_k \exp[j(\varphi_k + 2\pi f_d k T_s)] + N_k \tag{6.2}$$

where A_k presents the amplitude of the signal, f_d indicates the frequency deviation between the emitting laser and the local oscillator, T_s is the symbol period and N_k represents the amplitude spontaneous emission noise, which follows Gaussian distribution with a mean value of 0. φ_k is the phase noise caused by the laser linewidth and obeys the Wiener process, which is given by:

$$\varphi_k = \sum\nolimits_{i=-\infty}^{n} v_i \tag{6.3}$$

where v_i is an independent random variable following Gaussian distribution with a mean of 0, and the variance is presented as

$$\sigma_p^2 = 2\pi \Delta f T_s \tag{6.4}$$

where Δf presents the sum of the width of the emission laser and the local oscillator, T_s is the symbol period. From Eq. (6.4), it can be seen that the phase noise from the laser is accumulated gradually, which is related to the linewidth of laser and symbol period. Figure 6.11 shows the relationship between the phase noise and the number of symbols with the normalized linewidth $\Delta f \cdot \mathrm{Ts}$ of 10^{-6}, 10^{-5} *and* 10^{-4} respectively. It can be seen that the phase noise fluctuation increases when the linewidth of the laser increases. However, the phase noise obeys the Wiener process and changes slowly compared to the rapidly changing signal, so that the phase noise of a relatively short sequence can be regarded as the same.

The proposed carrier phase noise estimation algorithm consists of two steps. The first step uses the BPS algorithm to provide a rough estimate and a second one that provides a fine estimate. Suppose the algorithm finds the rough estimate range of phase noise $(\varphi_{min}, \varphi_{max})$ in the first step. Then divide the rough estimate range into several parts and define the FOD as:

$$d_k = min(|t - real\{y_k\}|) \tag{6.5}$$

where t is the real part of the standard constellation, such as $t = [-3, -1, 1, 3]$ for 16QAM signal. $Real\{y_k\}$ presents real part of the signal y_k. After collecting N symbols, the summation of first-order distance obtained by rotating phase signals can be expressed as

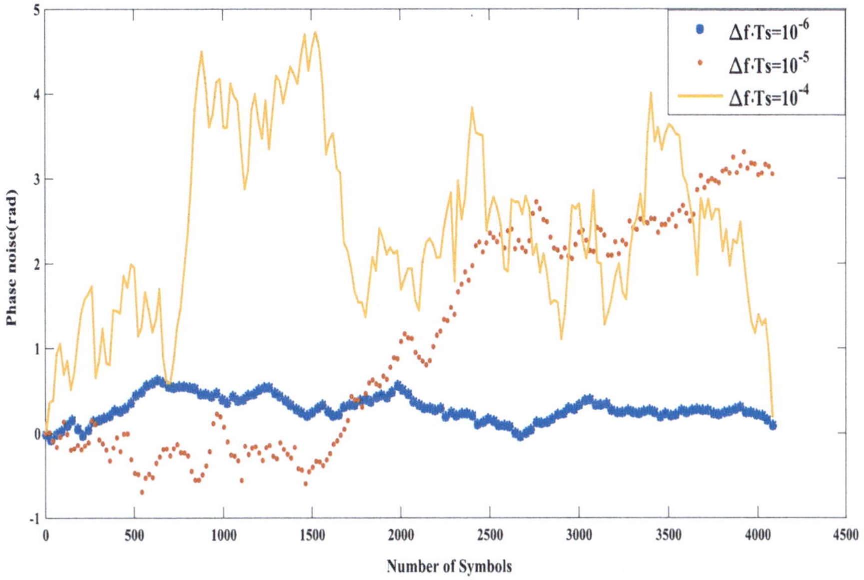

Fig. 6.11 The phase noise with the number of symbols

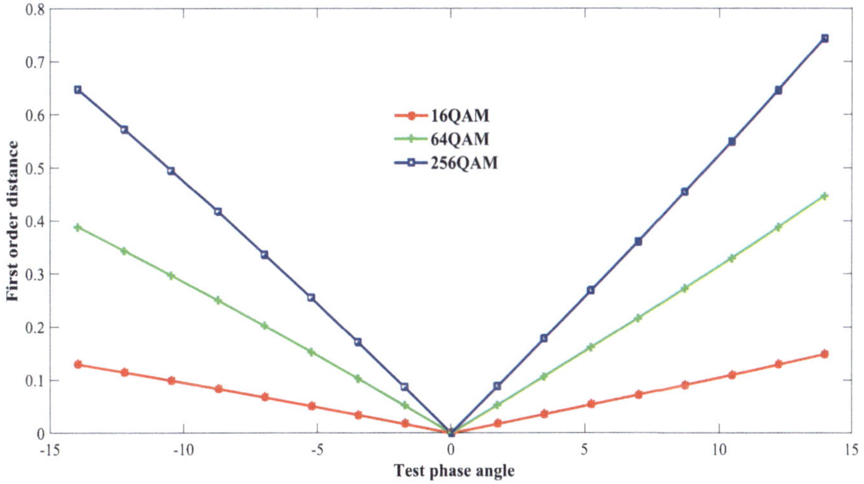

Fig. 6.12 The relationship of rotation phase with first order distance without phase noise loading

$$\mathbf{e}_b = \sum\nolimits_{k=1}^{N} d_k \tag{6.6}$$

Figure 6.12 depicts the first order distance d_k as a function of test angles φ_b for 16QAM, 64QAM and 256QAM with the phase noise of zero($\varphi_k = 0$). $\varphi_{error} = \varphi_k - \varphi_b$, if $\varphi_\mathrm{b} = 0$, $\mathrm{e}_b = \sum_{k=1}^{N} d_k$ is minimized. Therefore, we can use the relationship

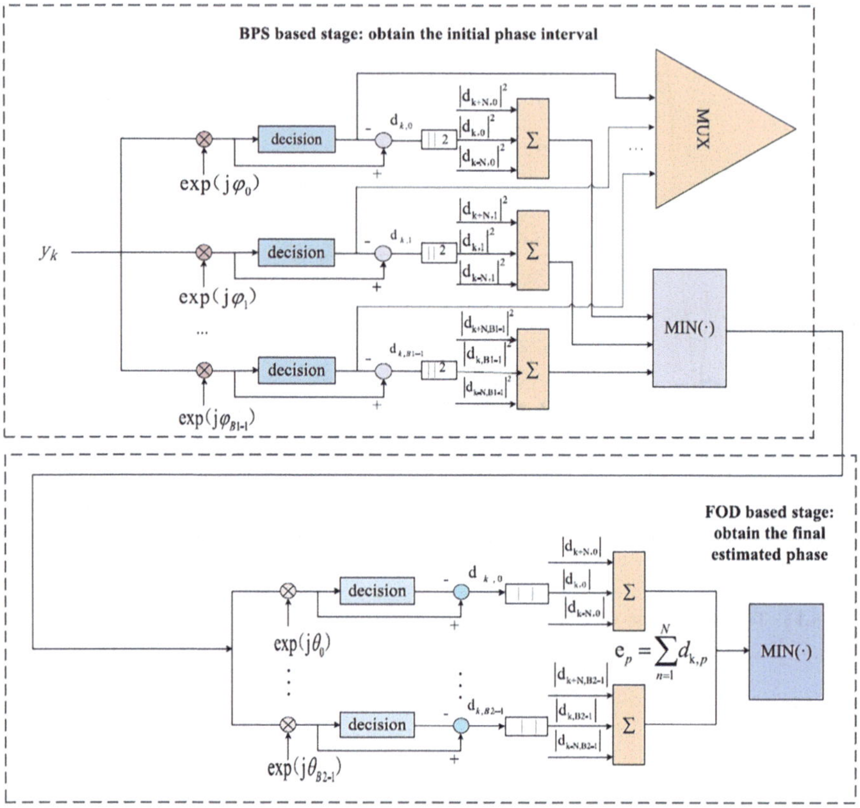

Fig. 6.13 Flowchart of the proposed algorithm

between the test phase and the first order distance to identify the optimum test phase angle for the kth sample.

The flowchart of the algorithm is described in Fig. 6.13. At the receiver side, the channel equalization including dispersion compensation, polarization mode dispersion compensation, frequency offset and clock recovery have been completed [57] afterword the symbol y_k is firstly rotated by B test phase angles φ_b:

$$\varphi_{\mathrm{b}} = \frac{b}{B} \cdot \frac{\pi}{2} - \frac{\pi}{4}, \varphi_{\mathrm{b}} \subset \left(-\frac{\pi}{4}, \frac{\pi}{4}\right), b = 0, 1, 2, \ldots B - 1 \tag{6.7}$$

The rotated symbols will be decided based on direct decision and the error value of each signal is calculated as:

$$\mathbf{d}_{k,b} = \left|y_k - [y_k]_D\right|^2 \tag{6.8}$$

where $\mathbf{ey_k} = \mathbf{y} * \mathbf{e}^{j*\varphi_b}$, k is the kth symbol and y is the received signal for phase recovery. $[y_k]_D$ represents the constellation point closest to the signal. After collecting N symbols, the error value can be expressed as:

$$\mathbf{e}_b = \sum\nolimits_{\mathbf{k}=1}^{N} d_{k,b} \tag{6.9}$$

In Eq. (6.9), according to the minimum e_b, we choose $f = sort(e_b)$ and select $f(1) = \varphi_{min}, f(2) = \varphi_{max}$. The rough estimate interval $(\varphi_{min}, \varphi_{max})$ is divided to obtain the test phase sequence θ_1, θ_2, $\ldots\theta p$. The rotated signal is presented as:

$$\mathbf{y}_{k,p} = \mathbf{y}_k \, \mathbf{exp}\left(\mathbf{j}\theta_{\mathbf{p}}\right) \tag{6.10}$$

Take the real part of the standard constellation point, the FOD of the rotated signal is:

$$\mathbf{d}_{\mathbf{k},\mathbf{p}} = min\left(\left|t - real\left\{y_{k,p}\right\}\right|\right) \tag{6.11}$$

After collecting N symbols, the summation of FOD obtained by rotating phase signals can be expressed as:

$$\mathrm{e}_p = \sum\nolimits_{n=1}^{N} d_{\mathrm{k},p} \tag{6.12}$$

Take the estimated phase value θ_{end} corresponding to the minimum FOD as the fine estimation of the phase noise, and the signal phase compensation can be presented as:

$$\mathbf{y}' = \mathbf{y}_k' \exp(-j\theta_{\mathrm{end}}) \tag{6.13}$$

where r is the recovered signal. In order to analyze the performance of the proposed algorithm, the coherent optical 16QAM transmission system is performed both in simulation and experimental environment. The parameters of the algorithm can be adjusted according to different modulation formats.

The complexity of the proposed algorithm is analyzed as well. The phase noise of *N* adjacent symbols can be considered to be the same. The first stage of the algorithm is used to determine the rough estimation range. Then the minimum FOD algorithm is used to find the exact value of the phase noise. It is assumed that the first stage rotates B1 phases and the second stage rotates B2. The algorithm uses multipliers, adders, comparators, and deciders to represent the complexity of algorithm. The complexity of the first stage contains 6NB1 + B1 multipliers, 6NB1 + B1 adders, 2B1 comparators, and NB1 deciders. The complexity of the second stage is analyzed as below.

1. Divide the rough estimation interval equally and obtain B2 rotation phases, which require B2 adders and B2 multipliers.

Table 6.2 The complexity comparison for different phase recovery algorithm

	Real multiplier	Real adder	Comparator	Decision
BPS	6NB + B	6NB + B	B	NB
ML	$6NB_M + 12N$	$6NB_M + B_M + 6N$	B_M	$NB_M + N$
FOD	6NB1 + B1 + B2 + 4NB2	6NB1 + B1 + B2 + 7NB2	2B1 + B2	NB1

2. Rotate B2 phase angles for each of the N symbols, as shown by Eq. (6.10), where the exponential operation can be converted into addition and multiplication operations, requiring 4NB2 multipliers and 2NB2 adders.
3. Calculate the FOD of each rotated signal, as demonstrated by Eq. (6.11). For 16QAM signals, one symbol requires 4 adders and 2 comparators. Accordingly, NB2 signals require 4NB2 adders and 2NB2 comparators.
4. Calculate the sum of FOD of the N symbols by rotating the same test phase, as shown in Eq. (6.12). One phase requires N adders; we have B2 phase, so that NB2 adders are required.
5. Compare the minimum FOD and the corresponding rotation test phase which is the fine phase noise, and the B2 rotation phases require B2 comparators.

Based on the above analysis, we obtain a total of $6NB1 + B1 + B2 + 4NB2$ multipliers based on the minimum FOD algorithm, $6NB1 + B1 + B2 + 7NB2$ adders, $2B1 + B2$ comparators and $NB1$ deciders, the total complexity can be presented as $O(12NB1 + 4B1 + 2B2 + 11NB2)$.

Table 6.2 shows the complexity comparison of the traditional BPS algorithm, ML and the minimum FOD method proposed in this section. Based on the characteristics of the mentioned phase noise distribution, the signal packets are processed in parallel, and the signals in the group are compensated by the same phase. N represents the number of symbols in the group and the calculated values of the complexity listed in Table 6.2. B is the number of test phases in the BPS method, and BM is the number of test phases required for the rough estimation of the first phase of the ML algorithm. B1 and B2 represent the number of rotation phases of the rough estimation phase and the fine estimation phase based on the first order distance.

In the simulation for 16QAM, N is 20 and B is 32 for the BPS algorithm [58]. B1 takes 16 and B2 takes 3 for FOD algorithm. Therefore, as for the traditional BPS scheme, the required number of multipliers/adders is 3872/3872 for 16QAM using such representative parameters, while the required number of multipliers/adders is 2179/2359 for 16QAM. Compared to the complexity of traditional BPS algorithm, the complexity of the proposed scheme is significantly reduced by the group factors of 1.77/1.64 (in the form of multipliers/adders) for 16QAM.

Simulation results: In order to evaluate the performance of the proposed FOD scheme, the coherent optical M-QAM transmission system is used as the simulation platform. In the simulation, we suppose that the frequency offset has been estimated and only the phase noise is considered. The laser phase noise is modeled as a Wiener process, shown in Eqs. (6.2)–(6.4). The signal to noise (SNR) is adjusted by loading the additive white Gaussian noise. And the performance of the algorithm is discussed by the parameters, including rotated phase number and linewidth tolerance.

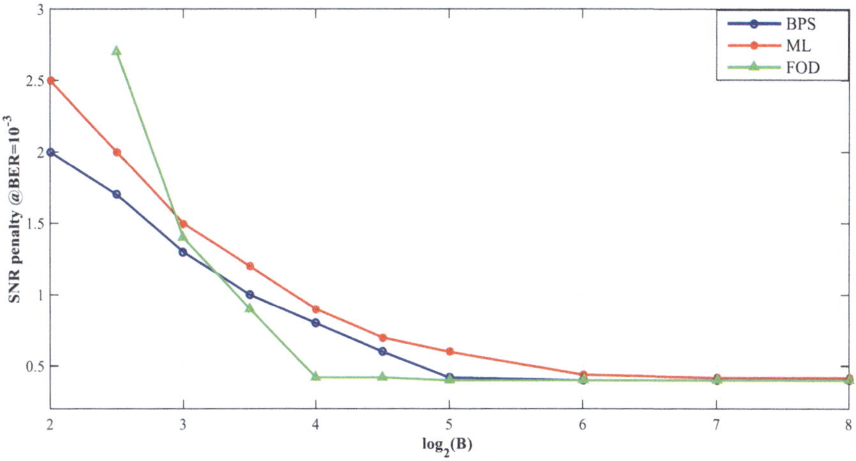

Fig. 6.14 The rel. of rotation no. with SNR penalty ($BER = 10^{-3}$) for 16QAM

Figure 6.14 shows the relationship of phase rotation number with SNR penalty for 16QAM using BPS, ML and the proposed FOD algorithms. As shown in Fig. 6.14, the SNR penalty of the three algorithms becomes smaller as the number of rotation phase increases. When the number of rotations is larger than 16 ($\log_2(B) > 4$), the minimum penalty for 16QAM is 0.38 dB, which is caused by encoding and quantization effects. Therefore, the rotation number of BPS is set to 32 ($\log_2(B) = 5$), and the rotation number of FOD in the first stage is set to 16 ($\log_2(B) = 4$) in the following simulation for 16QAM comparisons.

Figure 6.15 shows the same simulations for 64QAM and 256QAM. The number of phase rotations is larger than 64($\log_2(B) \geq 6$) for 64QAM using BPS and ML algorithm, while for FOD algorithm, $\log_2(B) \geq 5$ reaching the minimum penalty 0.4 dB. For 256QAM, form Eq. (6.7), which can be derived that once the rough estimate range of phase noise is small enough, the performance of the FOD algorithm will coincide with the BPS, ML algorithm when $\log_2(B) = 6$. The minimum penalty for 256QAM is 0.42 dB, which is caused by encoding and quantization effects.

In order to further evaluate the phase noise tolerance of individual algorithms, the relationship between BER and SNR is obtained for typical values of $\Delta f \cdot T_s$, Ts for 16QAM, 64QAM and 256QAM respectively in Fig. 6.16. The theoretical curve is given by [58]:

$$\mathrm{BER} = \left\{1 - \left(1 - \frac{2}{\log_2 M}\left(1 - \frac{1}{\sqrt{M}}\right)Q\left[\sqrt{\frac{3}{M-1}SNR}\right]\right)\right\} \cdot \left(1 + \frac{\log_2 M}{2(\sqrt{M}-1)}\right) \tag{6.14}$$

Experimental results: The performance of the proposed algorithm is experimentally demonstrated in 80 Gbit/s polarization multiplexing 16-QAM systems, as shown in Fig. 6.17.

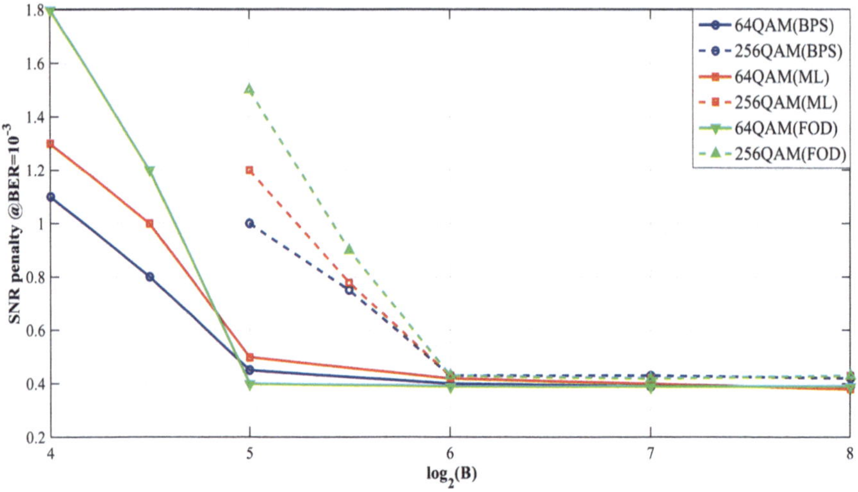

Fig. 6.15 The relationship of rotation number with SNR penalty ($BER = 10^{-3}$) for 64QAM and 256QAM

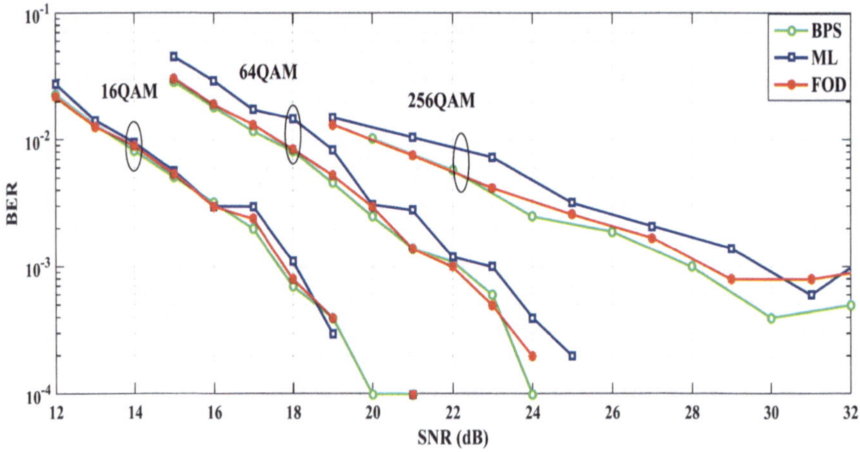

Fig. 6.16 Relationship between BER and SNR for 16QAM ($\Delta f \cdot T_s = 1 \cdot 10^{-4}$), 64QAM ($\Delta f \cdot T_s = 4 \cdot 10^{-5}$) and 256QAM ($\Delta f \cdot T_s = 8 \cdot 10^{-6}$)

At the transmitter, 10 Gbit/s binary pseudo-random bit sequence of length 2^{12} are obtained by a bit pattern generator, which is amplified, relatively delayed and combined together to generate two four-level signals. Then they are injected into the RF port of IQ modulator and generate a single polarization 16QAM signal. An external cavity laser with 100 kHz linewidth is used. Then the single polarization 16QAM signal is fed into a polarization multiplexer, which is split and recombined in orthogonal polarizations after delaying one of the split signals to de-correlate it from another split signal. The signal is transmitted through the fiber, and then

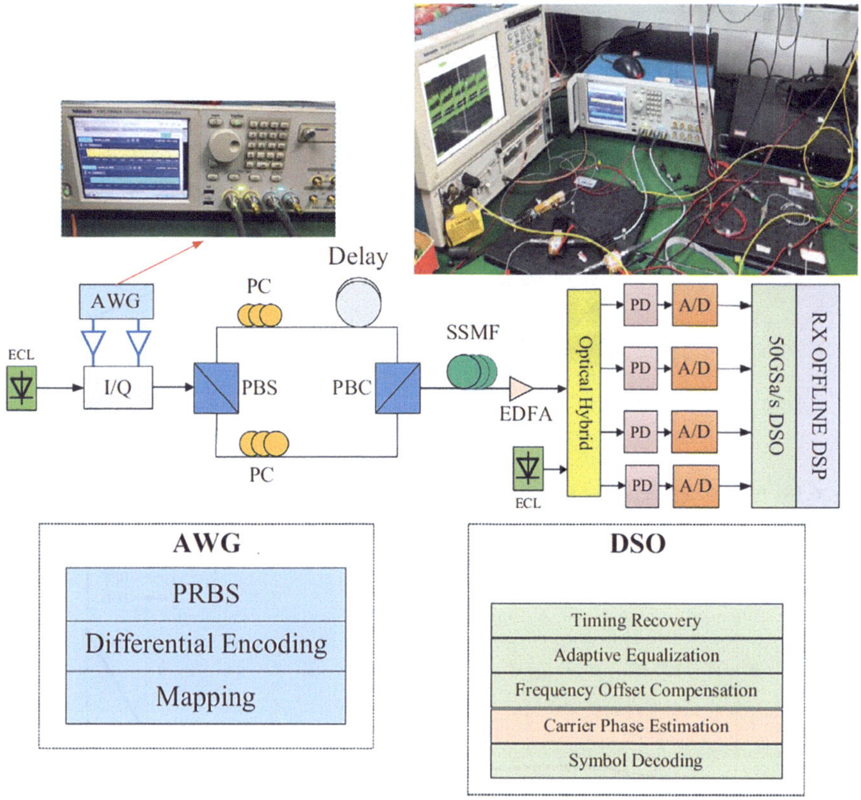

Fig. 6.17 Experimental Setup

amplified by the EDFA amplifies. The coherent receiver receives the signal and performs A/D sampling, IQ imbalance, time recovery, adaptive equalization, frequency offset compensation, and the proposed algorithm is used for carrier phase noise recovery [65].

Figure 6.18 shows the results for DP16QAM signals transmitted across 80 km SSMF, which indicates that the performance of the proposed algorithm is consistent with that of the original BPS in 80 km SSMF transmissions and the proposed FOD algorithm with much lower complexity. In order to further evaluate the phase noise tolerance of the algorithms for $BER = 10^{-3}$, Fig. 6.19 presents the phase noise tolerance for the FOD and BPS schemes, which shows the similar performance to laser phase noise. When $\Delta f \cdot T_s = 10^{-4}$, the OSNR required will increase sharply for both algorithms. It denotes that the proposed FOD algorithm can tolerate the normalized linewidth $\Delta f \cdot T_s$ of 10^{-4} for 16QAM signal.

Moreover, we have processed the experiment data (after channel equalization, and frequency offset estimation, the phase noise estimation being only considered) using BPS and FOD algorithm respectively. The data size is 151k, 300k, 450k, 749k

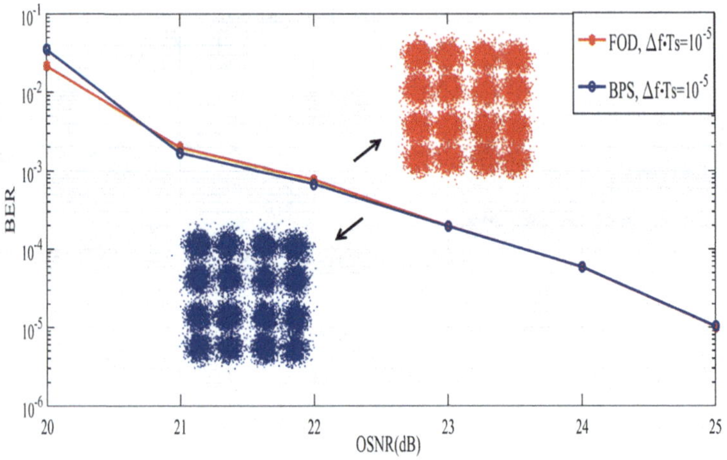

Fig. 6.18 BER as a function of OSNR for DP-16QAM signals after 80 km SSMF transmissions

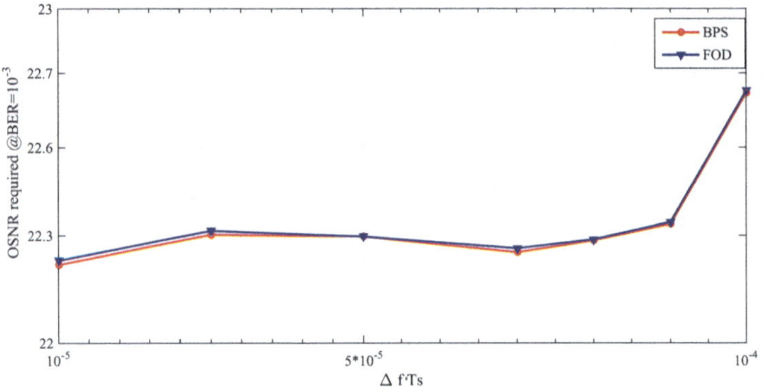

Fig. 6.19 Performance of phase noise tolerance for 16-QAM with BPS and FOD algorithm

and 1048k byte separately. The running time of central processing units (CPUs) with different algorithms is shown in Fig. 6.20, which shows that FOD algorithm can dramatically enhance time efficiency of the system and can save ~1.3–8.4 s for the group data.

In this section, a novel low complexity linewidth tolerant phase recovery algorithm based on first order distance is proposed for M-QAM, which is demonstrated experimentally in the 80 Gbit/s coherent transmission systems. The proposed FOD algorithm can tolerate the normalized linewidth $\Delta f \cdot T_s$ of 10^{-4} for 16QAM signal. It can immensely enhance time efficiency of the system and can save ~1.3–8.4 s for processing the group of 151k, 300k, 450k, 749k, and 1048k byte experimental data.

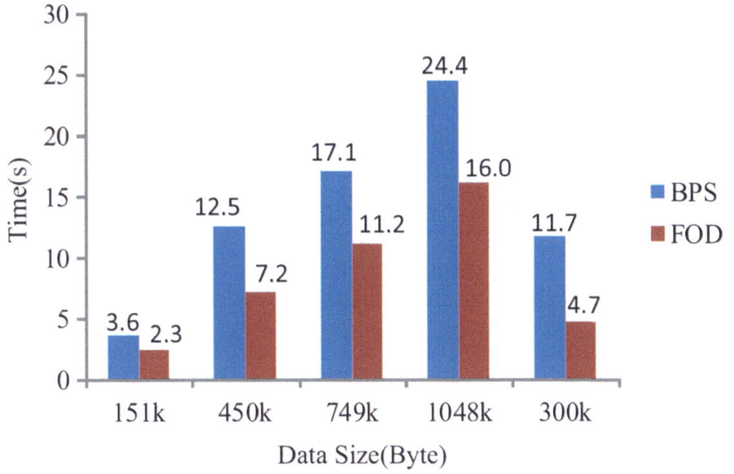

Fig. 6.20 The running time of CPUs with different algorithms

6.3.2 *Pilot-Aided Kalman Filter (PA-KF) CPR Algorithm for PS-64QAM*

In recent years, with the rapid development of various communication services such as distance education and intelligent transportation, the requirements for signal transmission are increasingly strict. The requirements for the bandwidth capacity of optical communication network also increases explosively. As a highly flexible technology that can improve the information transmission capacity, the probability shaping (PS) technology and the signal format combined with quadrature amplitude modulation (QAM) have received increasingly extensive attention [66, 67]. However, the PS signals are very sensitive to phase noise and have low noise tolerance. In order to improve the disadvantages of PS in coherent system, the research on CPR in digital signal processing (DSP) is essential. However, the unequal probability distribution in the constellation will bring difficulties to the DSP module at the receiver, especially for the CPR algorithm [68]. Several CPR schemes for probabilistic shaping signal formats have been confirmed [69–73]. The BPS algorithm has general mutual information (GMI) penalty with PS-QAM[1] signal under different shaping factors. The block length needs to be more than 300 symbols with high computational complexity to improve the penalty [68]. The quadrature phase shift keying (QPSK) segmentation algorithm based on constellation dependence is not suitable for phase recovery of high-order modulation format signals. The two-stage BPS algorithm has a long block length and low linewidth tolerance [70]. Recently, conventional Kalman filter (KF) for CPR in the coherent optical communication system has been proposed and demonstrated [74]. Although the Kalman filter has high accuracy for CPR, it has an intensive computational complexity. Therefore, the high tolerant CPR scheme with low complexity is badly needed for the PS-QAM system.

In this section, we propose a CPR scheme based on pilot aided Kalman filter (PA-KF) in PS-64QAM system. The block of PA-KF algorithm can reduce the complexity. These symbols in a block can be updated with only Kalman gain and state value of the pilot symbol to reduce the amount of calculation. Simulation result shows that the proposed scheme has better linewidth tolerance than BPS. The result shows the complexity of the PA-KF is reduced by 99% and the GMI of PA-KF CPR increases by ~1.18 bit/symbol compared with BPS.

Assuming that all channel impairments except phase noise are compensated, the received signal at discrete time n is given by:

$$r_n = a_n \exp(j\theta_n) + \eta_n \tag{6.15}$$

where a_n is the n_{th} transmitted symbol. η_n models the collective amplified emission (ASE) noise following complex circularly symmetric Gaussian random process, and θ_n is the laser noise modeled as a Wiener process.

$$\theta_n = \sum\nolimits_{i=-\infty}^{n} \nu_i \tag{6.16}$$

v_n is a Gaussian random variable with zero mean and variance σ_v^2 [73]:

$$\sigma_v^2 = 2\pi(\Delta f \cdot T_s) \tag{6.17}$$

T_s is the symbol period and Δf is the sum of the linewidths of the Tx laser and the local oscillator.

In order to compensate the CPN, the state vector can be designed as follows:

$$x_k = \exp(j\theta_k) \tag{6.18}$$

The measurement vector is established the as shown in equations:

$$z_k = d_k \cdot \widehat{x}_{k|k-1} \tag{6.19}$$

The measurement vector is the result of multiplying the d_k by the state vector. $\widehat{x}_{k|k-1 \cdot d_k}$ is the standard constellation point determined by the receiving signal compensation when the Euclidean distance is minimum:

$$d_k = dec\left\{ \min_{i=1-64} \left| r_k \cdot x_k \widehat{|k-1} - d_l \right| \right\} \tag{6.20}$$

The innovation vector can be written as equations:

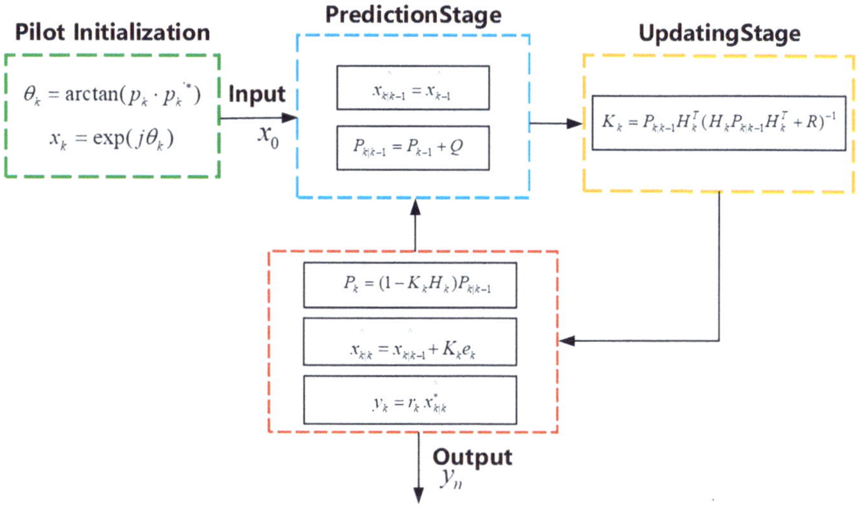

Fig. 6.21 The structure of the PA-KF algorithm

$$e_k = r_k - z_k \tag{6.21}$$

The observation matrix in the system is described as:

$$H_k = d_k \tag{6.22}$$

In the PA-KF CPR scheme, the signals consist of an initial pilot sequence and a periodic payload. The inserted pilot symbols are used for continuous phase tracking and KF phase estimate. In the algorithm, the current vector is updated by the measure value of the previous time as the predicted value of the current time. The measurement update calculation is carried out through the observation equation.

The algorithm can be divided into the pilot initialization, prediction stage and the updating stage. The structure of PAKF algorithm is shown in Fig. 6.21.

For the PA-KF CPR scheme, the initial state vector x_k of each KF block is calculated from the estimated phase of the QPSK pilot signal in the pilot initialization stage. There are p pilot symbols with QPSK inserted into the N symbols payload sequence at the receiver, and the insertion ratio is p/N. The prediction stage has two steps: (1) predicting the state vector $\widehat{X_{k|k-1}}$ by the previous symbol $\widehat{X}_{k-1}$; (2) predicting the priori error covariance $P_{k|k-1}$ by the prediction error covariance of the previous symbol P_{k-1}. The prediction process is described as:

$$\widehat{X_{k|k-1}}' = F_{k-1}\widehat{X'_{k-1}} \tag{6.23}$$

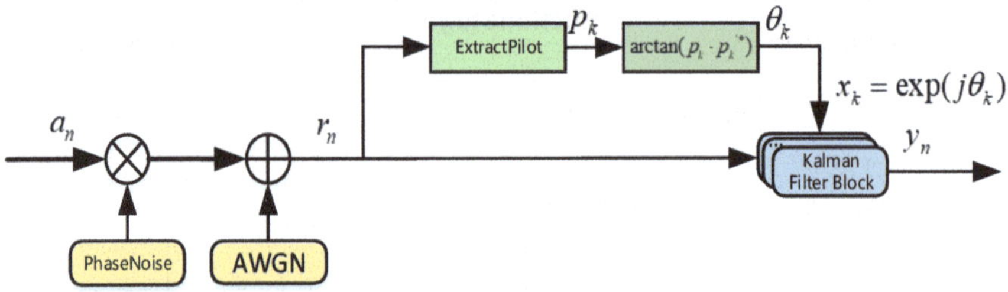

Fig. 6.22 The schematic of the PA-KF CPR

$$P_{k|k-1} = F_{k-1}P_{k-1}F_{k-1}^T + Q \tag{6.24}$$

Where $\widehat{X_{k|k-1}}$ is the state vector at time k is estimated by time k-1. F_{k-1} is the state transition matrix and Q is the process noise covariance.

The updating stage has three steps: (1) calculating the Kalman Gain; (2) calculating $\widehat{X_{k|k}}$ according to the innovation vector and the priori state vector $\widehat{X_{k|k}}$; (3) updating the error covariance P_k. The correction process is described as:

$$K_k\& = P_{k|k-1}H_k^T\left(H_kP_{k|k-1}H_k^T + R\right)^{-1} \tag{6.25}$$

$$\widehat{X_{k|k}}\& = \widehat{X_{k|k-1}} + K_k e_k \tag{6.26}$$

$$P_k\& = (I - K_kH_k)P_{k|k-1} \tag{6.27}$$

Where R is measurement noise covariance. There is a certain correlation between the optimal bit error performance with the value of Q and R. The value of Q is related to the laser linewidth and R is bound up with white Gaussian white noise in the system. In this scheme, R and Q are given as R = 10 and Q = 1e-4.

The PA-KF CPR scheme is shown in Fig. 6.22. Except for phase noise, all the channel impairments are compensated in the system. Firstly, the PS-64QAM signal is divided into k blocks with n transmission symbols in each block. The K pilot symbols in transmission signal are extracted which are conjugate multiplied with the original pilot symbols to obtain the pilot phase noise θ_k. x_k is set as the initial state vector of each parallel KF block to improve the calculation rate and reduce the complexity. Finally, r_n is compensated by the estimated state vector to realize accurate phase recovery.

This section mainly analyzes the effectiveness and complexity of the proposed PA-KF CPR scheme. In PS systems, the GMI will suddenly decline at the optimal λ [68] when using the BPS algorithm for phase recovery at low SNR. The BPS algorithm has a strong dependence on PS, even for long noise rejection windows. The impact of different PS factors on GMI in the PA-KF and BPS scheme is evaluated.

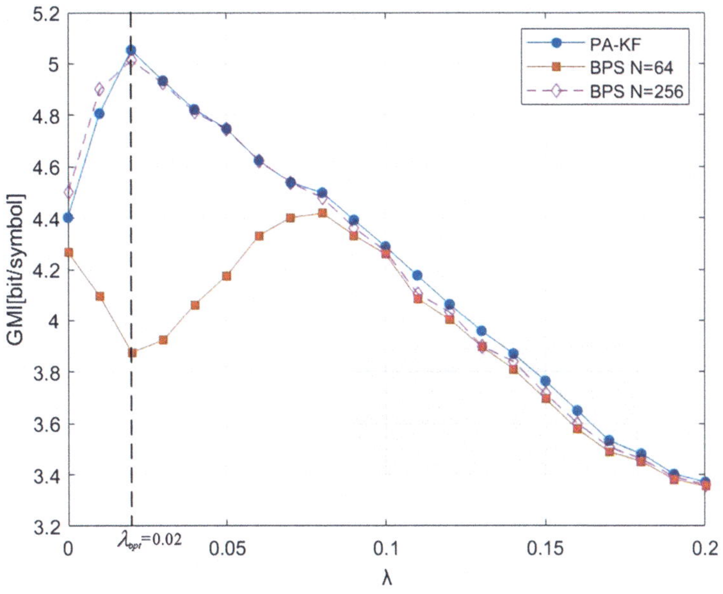

Fig. 6.23 GMI results with different phase recovery algorithms in PS-64QAM systems

Figure 6.23 shows the GMI performance of PA-KF CPR and BPS (N = 64, N = 256) in different PS factors λ. The SNR is 17 dB. The sudden drop of GMI under BPS (N = 64) is obvious in λ = 0.02, yet the PA-KF CPR does not cause a sudden drop at different PS factor. The GMI of PAKF CPR increases by ~1.18 bit/symbol compared with that of BPS (N = 64). When using the BPS algorithm, the window length of system is important. The BPS (N = 256) scheme has similar GMI with PA-KF scheme at λ = 0.02. Although the penalties imposed by PS on the BPS can be eventually mitigated by extremely long windows length, which seriously increases the computational complexity.

Figure 6.24 gives the simulated bit-error ratio (BER) for different phase recover algorithm in a range of laser linewidth. The BPS method uses equivalent 64 test-phase angle and the PA-KF algorithm. N is the block length of the single-stage BPS algorithm. It can be observed that the PAKF has better BER performance compared with BPS methods.

As is shown in Table 6.3, the computational complexity of PA-KF and BPS algorithms are compared. For PA-KF, it contains 108 + 2 K real multipliers, 86 real adders, 34 comparators, and 34 decisions. For BPS, the window length N = 64\256 and the number of test phase B = 32\64. The BPS requires 12NB + 2B real multiplier, 12NB + 2B real adders, 2B comparison, and 2NB decision. Compared to the BPS algorithm, the complexity of the PA-KF algorithm is lower than that of the BPS algorithm and it reduces the complexity by about 99.88% (N = 256 B = 64) and 99.04% (N = 64 B = 32).

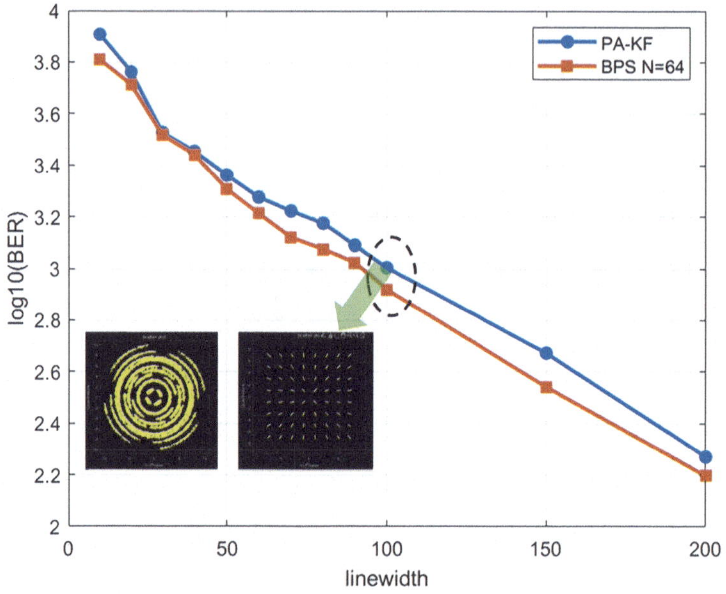

Fig. 6.24 BER against linewidth under PS-64QAM for different phase recovery algorithm with SNR of 17 dB

Table 6.3 Computation complexity

	Real multiplier	Real adder	Comparison	Decision
PA-KF	108 + 2 K	86	34	34
BPS	12NB + 2B	12NB + 2B	2B	2NB

In this section, we proposed the PA-KF carrier phase recovery algorithm for PS-64QAM at different linewidth. This scheme has better CPR performance and no penalty with PS signal at different shaping factors with low complexity. The result shows the complexity of the PA-KF is reduced by 99% and the GMI of PA-KF CPR increases by ~1.18 bit/symbol compared with BPS.

6.4 Summary

In this chapter, we propose some novel equalization and compensation algorithms for optical fiber communication systems. Firstly, a joint CMA-DDLMS algorithm enabled a 2 × 20G Baud PAM4 transmission over 20 km, reducing BER to 1.0×10^{-3} at 0 dB received power and analyzing relevant factors' impacts. Secondly, the proposed Jones-Volterra training sequence (JVTS) mitigated nonlinearities and crosstalk in FMF systems, improving received power and BER. Finally, a feed-forward carrier phase estimation algorithm with FOD approximation lowered complexity while guaranteeing accuracy.

References

1. T. Mori, T. Sakamoto, M. Wada, et al., Few-mode fibers supporting more than two LP modes for mode-division-multiplexed transmission with MIMO DSP. J. Lightwave Technol. **32**(14), 2468–2479 (2014)
2. Y. Chen, A. Lobato, et al., 41.6 Tbit/s C-band SDM OFDM transmission through 12 spatial and polarization modes over 74.17 km few mode fiber. J. Lightwave Technol. **33**, 1440–1444 (2015)
3. G. Rademacher, R. Ryf, H. Chen, et al., Long-haul transmission over few-mode fibers with space-division multiplexing. J. Lightwave Technol. **36**(6), 1382–1388 (2018)
4. J. Li, C. Cai, J. Du, et al., Ultra-low-noise mode-division multiplexed WDM transmission over 100-km FMF based on a second-order few-mode Raman amplifier. J. Lightwave Technol. **99**, 1–3260 (2018)
5. G. Rademacher, R.S. Luis, B.J. Puttnam, et al., High capacity transmission with few-mode fibers. J. Lightwave Technol. **37**(2), 425–432 (2018)
6. S. Jiang, C. Liang, L. Ma, et al., Ultra-low-loss broadband all-fiber mode selective couplers for MIMO-less MDM transmission. J. Lightwave Technol. **38**(8), 2376–2382 (2020)
7. K. Shibahara, T. Mizuno, Y. Miyamoto, Long-haul mode multiplexing transmission enhanced by interference cancellation technique, in *45th European Conference on Optical Communication (ECOC 2019)*, (2019)
8. A. Li, X. Chen, A.A. Amin, et al., Space-division multiplexed high-speed superchannel transmission over few-mode fiber. J. Lightwave Technol. **30**(24), 3953–3964 (2012)
9. Z. Mohammadzahery, M. Jandaghi, S. Alipour, S. Salimian Rizi, E. Hajinia, E. Aghayari, H. Nabavi, Nonlinear spatial reshaping of pulsed beam in a step-index few-mode optical fiber. Opt. Express **29**, 10716–10725 (2021)
10. F. Liu, W. Zhang, W. Ping, Z. He, Fault detection sensitivity enhancement based on high-order spatial mode trend filtering for few-mode fiber link. Opt. Express **29**, 5226–5235 (2021)
11. R.S. Luís et al., 1.2 Pb/s throughput transmission using a 160 μm cladding, 4-core, 3-mode fiber. J. Lightwave Technol. **37**, 1798–1804 (2019)
12. B.J. Puttnam et al., High data-rate and long distance MCF transmission with 19-core C+L band cladding-pumped EDFA. J. Lightwave Technol. **38**, 123–130 (2020)
13. K. Shibahara, T. Mizuno, D. Lee, Y. Miyamoto, Advanced MIMO signal processing techniques enabling long-haul dense SDM transmissions. J. Lightwave Technol. **36**, 336–348 (2018)
14. G. Rademacher et al., Mode-dependent crosstalk penalty in few-mode multi-core fiber transmission, in *2019 Conference on Lasers and Electro-Optics Europe and European Quantum Electronics Conference, OSA Technical Digest (Optical Society of America, 2019), paper ci_2_2*, (2019)
15. T. Mori et al., Strongly-coupled five-mode ring-core fiber for MDM transmission with MIMO DSP, in *Optical Fiber Communication Conference, OSA Technical Digest*, (Optical Society of America, 2017) paper Tu2J.3
16. S.A. Sheikh, P. Fan, Improved square contour algorithm for blind equalization of QAM signals, in *2006 IEEE International multitopic conference, Islamabad*, (2006), pp. 116–119
17. Y. Wang, L. Tao, Y. Wang, N. Chi, High speed WDM VLC system based on multi-band CAP64 with weighted pre-equalization and modified CMMA based post-equalization. IEEE Commun. Lett. **18**(10), 1719–1722 (2014b)

18. R. Rumipamba-Zambrano, J. Perelló, S. Spadaro, Route, modulation format, MIMO, and spectrum assignment in flex-grid/MCF transparent optical core networks. J. Lightwave Technol. **36**, 3534–3546 (2018)
19. B. Liu, L. Zhang, X. Xin, A. Novel, MIMO DSP based on matrix transformation for joint few-mode/multi-core optical transmission system, in *Asia Communications and Photonics Conference 2015, OSA Technical Digest (Online)*, (Optical Society of America, 2015), p. paper ASu5D.4
20. P. Sillard, M. Astruc, D. Boivin, H. Maerten, L. Provost, Few-mode fiber for uncoupled mode-division multiplexing transmissions, in *Proceedings of the 37th European Conference Exposition on Optical Communications*. Paper Tu.5.LeCervin, vol. 7, (2014)
21. C. Koebele, M. Salsi, D. Sperti, et al., Two-mode transmission with digital inter-modal cross-talk mitigation, in *ECOC Technical Digest*, (Optica Publishing Group, 2011)
22. Y. Zhao, Z. Tao, H. Li, X. Su, S. Oda, Y. Aoki, T. Hoshida, Channel spacing monitor based on periodic training sequence in DWDM system. J. Lightwave Technol. **35**, 1422–1428 (2017)
23. M. Gagni, F.P. Guiomar, S. Wabnitz, A.N. Pinto, Simplified high-order Volterra series transfer function for optical transmission links. Opt. Express **25**, 2446–2459 (2017)
24. V. Vgenopoulou, N.P. Diamantopoulos, I. Roudas, S. Sygletos, IEEE, MIMO nonlinear equalizer based on inverse Volterra series transfer function for coherent SDM systems, in *Optical Fiber Communication Conference (OFC)*, (Optica Publishing Group, 2019)
25. J.H. Ke, K.P. Zhong, Y. Gao, A. Bakhshali, J.C. Cartledge, Low-complexity joint frequency offset and carrier phase estimation based on QPSK partitioning for DP 16-QAM, in *Optical Fiber Communication Conference*, (Optica Publishing Group, 2013), p. OTu3I.5
26. A.P.T. Lau, Y. Gao, Q. Sui, D. Wang, Q. Zhuge, M. Morsy-Osman, et al., Advanced DSP techniques enabling high spectral efficiency and flexible transmissions: Toward elastic optical networks. IEEE Signal. Process. Mag. **31**(2), 82–92 (2014)
27. R.W. Tkach, Scaling optical communications for the next decade and beyond. Bell Labs Tech. J. **14**(4), 3–9 (2010)
28. P. Winzer, High-spectral-efficiency optical modulation formats. J. Lightwave Technol. **30**(8), 3824–3835 (2012)
29. L. Kumar, V. Sharma, A. Singh, Feasibility and modelling for convergence of optical-wireless network—A review. AEU Int. J. Electron. Commun. **80**, 144–156 (2017)
30. A.R. Barroso, J. Johnson, Optical wireless communications omnidirectional receivers for vehicular communications. AEU Int. J. Electron. Commun. **79**, 102–109 (2017)
31. Y.S. Li, Y.Y. Wang, T. Jiang, Norm-adaption penalized least mean square/fourth algorithm for sparse channel estimation. Signal. Process. **128**, 243–251 (2016)
32. Z.N. Tao, L. Li, A. Isomura, T. Hoshida, J.C. Rasmussen, Multiplier-free phase recovery for optical coherent receivers, in *Optical Fiber Communication Conference*, (Optica Publishing Group, 2008), pp. 1–3
33. Y.S. Li, Y.Y. Wang, T. Jiang, Sparse least mean mixed-norm adaptive filtering algorithms for sparse channel estimation applications. Int. J. Commun. Syst. **30**(8), e3181 (2017)
34. Y.S. Li, Y.Y. Wang, T. Jiang, Sparse-aware set-membership NLMS algorithms and their application for sparse channel estimation and echo cancelation. AEU Int. J. Electron. Commun. **70**(7), 895–902 (2016)
35. A. Tarighat, R.C. Hsu, A.H. Sayed, B. Jalali, Digital adaptive phase noise reduction in coherent optical links. J. Lightwave Technol. **24**(3), 1269–1276 (2006)
36. H. Louchet, K. Kuzmin, A. Richter, Improved DSP algorithms for coherent 16-QAM transmission, in *Proceedings of the ECOC'08, Brussels, Belgium, Sep. 21–25*, (IEEE, 2008), p. Tu.1.E.6
37. A. Viterbi, Nonlinear estimation of PSK-modulated carrier phase with application to burst digital transmission. IEEE Trans. Inf. Theory **29**(4), 543–551 (1983)
38. T. Pfau, S. Hoffmann, R. Noe, Hardware-efficient coherent digital receiver concept with carrier recovery for M-QAM constellations. J. Lightwave Technol. **27**(8), 989–999 (2009)

39. X. Su et al., Laser-linewidth-tolerant and low-complexity two-stage carrier phase estimation for M-QAM with coherent detection, in *Asia Communications and Photonics Conference*, (Optica Publishing Group, 2013)
40. S.M. Bilal, G. Bosco, P. Poggiolini, et al., Low-complexity linewidth-tolerant carrier phase estimation for 64-QAM systems based on constellation transformation, in *39th European Conference and Exhibition on Optical Communication (ECOC 2013)*, (IET, Stevenage UK, 2013), pp. 1–3
41. S.M. Bilal, G. Bosco, J. Cheng, et al., Performance and complexity comparison of CPE algorithms for 256-QAM optical signals, in *Optical Fiber Communication Conference*, (Optica Publishing Group, 2015), p. W1E-6
42. S.M. Bilal, C. Fludger, G. Bosco, Multi-stage CPE algorithms for 64-QAM constellation partitioning, in *Optical Fiber Communication Conference*, (Optical Society of America, 2014), p. M2A.8
43. S.M. Bilal, C. Fludger, V. Curri, et al., Multistage carrier phase estimation algorithms for phase noise mitigation in 64-quadrature amplitude modulation optical systems. J. Lightwave Technol. **32**(17), 2973–2979 (2014)
44. S.M. Bilal, G. Bosco, J. Cheng, et al., Carrier phase estimation through the rotation algorithm for 64-QAM optical systems. J. Lightwave Technol. **33**(9), 1766–1773 (2015)
45. M. Xiang, S.N. Fu, L. Deng, et al., Low-complexity feed-forward carrier phase estimation for M-ary QAM based on phase search acceleration by quadratic approximation. Opt. Express **23**(15), 19142–19153 (2015)
46. F. Rottenberg, T.H. Nguyen, ML and MAP phase noise estimators for optical fiber FBMC-OQAM systems, in *IEEE International Conference on Communications*, (2017), pp. 1–6
47. Y. Li, Z. Jiang, W. Shi, X. Han, B. Chen, Blocked maximum correntropy criterion algorithm for cluster-sparse system identifications. IEEE Trans. Circ. Syst. II Express Briefs **66**(11), 1915–1919 (2019)
48. Y. Li et al., Mixed norm constrained sparse APA algorithm for satellite and network echo channel estimation. IEEE Access **6**, 65901–65908 (2018)
49. F. Buchali, F. Steiner, G. Böcherer, L. Schmalen, P. Schulte, W. Idler, Rate adaptation and reach increase by probabilistically shaped 64-QAM: An experimental demonstration. J. Lightw. Technol. **34**(7), 1599–1609 (2016)
50. D. Pilori, L. Bertignono, A. Nespola, et al., Comparison of probabilistically shaped 64QAM with lower cardinality uniform constellations in long-haul optical systems. J. Lightwave Technol. **36**(2), 501–509 (2018)
51. D.A. Mello, F.A. Barbosa, J.D. Reis, Interplay of probabilistic shaping and the blind phase search algorithm. J. Lightwave Technol. **36**, 5096–5105 (2018)
52. A. Jain, A.I.A. El-Rahman, J.C. Cartledge, Optimization of two-stage carrier phase estimation for probabilistic shaping, in *45th European Conference on Optical Communication*, (2019), pp. 1–4
53. J. Hu, Z. Sun, X. Xu, Y. Lu, Y. Qiao, Carrier phase recovery exploiting recursive blind phase search algorithm for probabilistic shaping 64QAM systems, in *2020 Asia Communications and Photonics Conference and International Conference on Information Photonics and Optical Communications*, (2020), pp. 1–3
54. Z. Lin, Q. Xiang, Y. Yang, Y. Yao, Robust and linewidth tolerant two-stage carrier recovery scheme for probabilistically shaped QAM system, in *2020 Conference on Lasers and Electro-Optics Pacific Rim*, (2020), pp. 1–2
55. Q. Zhang, C. Shu, Viterbi and Viterbi algorithm based phase recovery for probabilistically shaped signals. J. Lightwave Technol. **39**(5), 1364–1370 (2021)
56. M. Morsy-Osman, Q. Zhuge, L.R. Chen, D.V. Plant, Feedforward carrier recovery via pilot-aided transmission for single carrier systems with arbitrary M-QAM constellations. Opt. Exp. **19**(24), 24331–24343 (2011)

57. F. Liu, Y. Lin, L.P. Barry, Y. Yu, Kalman filter for carrier phase recovery in coherent transmission systems employing fast tunable lasers, in *Asia Communications and Photonics Conference*, (2017), pp. 1–3
58. E. Ip, J.M. Kahn, Feedforward carrier recovery for coherent optical communications. J. Lightwave Technol. **25**(9), 2675–2692 (2007)

Chapter 7
Conclusions and Future Challenges

7.1 Conclusions and Future Challenges

7.1.1 Conclusions

In this book, starting from the technical principles commonly used in high-speed large capacity fiber optical communication, the basic knowledge and principles of multidimensional modulation and coding technology, multi-carrier generation and transmission technology, multi-band fiber transmission technology, space division multiplexing technology, and equalization and compensation algorithms are introduced. These technologies are innovatively applied to the system, thus significantly improving the performance of communication in terms of capacity, quality, reliability, and other aspects.

By utilizing multidimensional modulation and coding technologies, we proposed and implemented various schemes, including a novel adaptive non-uniform PSK modulation scheme, a 3D-AH-8/64QAM scheme, probabilistic-shaping four-dimensional modulation, and ATTCM-based concatenated coded modulation within a GFDM system.

For the research of multi-carrier generation and transmission technology, we used an SSB modulator-based RFS to generate 50 frequency—locked optical carriers with a pitch of 12.5 GHz, featuring remarkable flatness and stability. Subsequently, a new dual—band RFS architecture based on a DP—MZM is put forward, which doubles efficiency and cuts costs in half. Finally, by using RFS the generated carriers (TNR $>$ 25 dB), a 2.56—Tb/s PDM—RZ—16QAM CO—WDM system is experimentally verified over 800 km of standard single-mode fiber, attaining a spectral efficiency of 5.1 b/s/Hz.

To advance extended wavelength division multiplexing technology, we implemented a faster-than-Nyquist scheme in multi-band transmission systems and proposed an optimization scheme for compensating the spectral power tilt caused by SRS in the S + C + L band system. And we also achieved the ultra-wideband

F. Tian, X. Xin, *High-Speed Large Capacity Optical Fiber Communications*,
Wireless Networks, https://doi.org/10.1007/978-3-031-94211-2_7

transmission of 480, 50 GHz spaced channels by using PDM 256-QAM modulation over 20 km of SSMF.

For space division multiplexing technology, we introduced the principle and scheme of the SDM transmission, we proposed a TA-CMA algorithm to mitigate signal impairments and a VSS-FDLMS algorithm, applying it to a 7-core 6-mode optical SDM transmission system. We demonstrated the experiment of a 261.7 Gbit/s 16-QAM signal transmitting a 2.5 km seven-core fiber with a KK receiver and proposed XT-MIMO-Volterra to reduce inter-core crosstalk and nonlinear degradation. We also proposed a SCHD scheme based on 4D modulation, which achieves a maximum rate of 165.3 Gbit/s.

Through the study of advanced equalization and compensation technology, we proposed and implemented new channel equalization and compensation algorithms such as Joint algorithm composed of CMA, DDLMS and SCA-CMMA-based MIMO equalization, as well as the JVTS nonlinear compensation algorithms, and two carrier phase recovery algorithms to recover the original signal and improve the accuracy and reliability of information transmission.

7.2 Future Challenges

At present, various achievements have been made in the research of high-speed large capacity systems for fiber optic communication, ensuring the communication quality of the system. But for fiber optic communication, there are still upcoming issues waiting to be resolved. For example, accessing optical networks and the security performance after accessing optical networks have become new challenges faced by high-speed large capacity fiber optic communication.

7.2.1 Optical Access Network

Optical access network, also known as "fiber optic user loop", refers to the collection of access connections that use or partially use optical transmission between business nodes or remote modules and user devices, sharing the same network side interface. It is a fiber optic transmission system designed for access network environments. It mainly consists of Optical Line Terminal (OLT), Optical Distribution Network (ODN), Optical Network Unit (ONU), and Adaptation Function (AF) modules. The OAN of an optical distribution network (ODN) composed of passive optical components such as passive optical splitters is called passive optical network (PON), while the OAN of an ODN composed of electrical multiplexing technology is called active optical network (AON).

The current active optical access network mainly refers to the integrated digital loop carrier system (IDLC). Active optical networks use electric multiplexers for branching, which can extend transmission distance and have the characteristics of simple technology, easy implementation, and strong networking ability. Passive optical networks can minimize the cost of fiber optic loops. Due to the enormous capacity of fibers, they have higher efficiency and development potential. Moreover, owing to the use of optical splitters for branching, they are easy to expand business with low maintenance costs.

With the advancement of technology and the increasing volume of IP services in the network, optical access network technology has become the key to the development of the entire network. The development trend of access network technology is crucial to the development of the network. At present, the bandwidth, ATM, and hybrid networking of optical access network technology are basically where the network is heading.

7.2.2 *Optical Network Security*

In recent years, fiber optic communication systems have rapidly developed with higher speed and larger capacity, bringing great convenience to people's daily lives. Optical network technology is also widely used in backbone transmission, urban area networks, access networks, data center optical interconnection and other scenarios thanks to its advantages of large transmission capacity, low loss, and free of electromagnetic interference. As the most fundamental physical infrastructure for information and communication, one of its characteristics is its wide coverage and high susceptibility to eavesdropping and intrusion. Therefore, the issue of transmission security has gradually grabbed people's attention, and how to achieve secure transmission of fiber optic transmission systems has become a recent hot topic, and optical network security transmission technology is also developing slowly.

There are currently a large number of security protection solutions for optical networks, such as Optical Code Division Multiple Access (OCDMA), Quantum Cryptography, Chaotic Encryption, and Photonic Firewall. By taking the advantage of the characteristics of large-scale networking and interconnection in optical networks, secure data transmission can also be achieved by continuously changing the transmission path of services, adjusting the modulation format used by services, and changing the spectrum resources allocated to services.

But with the development of intrusion technology, even the safest protection methods will always be breached. No protection technology is once and for all absolutely safe. As a result, the security issue of optical networks remains a major challenge we face.

7.3 Summary

In this chapter, we summarize the key technologies of high-speed large capacity optical fiber communication and analyze the future challenges.

The manufacturer's authorised representative in the EU is Springer Nature Customer Service Centre GmbH, Europaplatz 3, 69115 Heidelberg, Germany. If you have any concerns regarding our products, please contact ProductSafety@springernature.com

Printed and bound by CPI Group (UK) Ltd, Croydon, CR0 4YY
30/09/2026
02225590-0001